Surface Color Perception

Alternate color perceptions. The area of intersection of each pair of circles is the result of combining paints of the two colors in equal amounts. In each, the intersection area can be perceived as either the mixture color when the circles are seen in the same plane or as the color of one circle through the transparent color of the other circle when the circles are seen in different planes.

Surface Color Perception

JACOB BECK

CORNELL UNIVERSITY PRESS | Ithaca and London

First published 1972 by Cornell University Press.
Published in the United Kingdom by Cornell University Press Ltd., 2-4 Brook Street, London W1Y 1AA.

International Standard Book Number 0-8014-0704-4
Library of Congress Catalog Card Number 76-38118

Printed in the United States of America by Vail-Ballou Press, Inc.

Librarians: Library of Congress cataloging information appears on the last page of the book.

To the memory of
Robert B. MacLeod

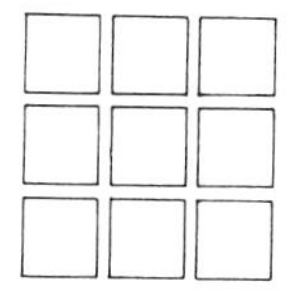

Preface

This book is about how people see surface color. By surface color I mean color that is perceived as an attribute of a surface. Experimental studies of surface color perception have centered on such questions as the following: How do the processes of contrast and adaptation affect the perception of a surface color? What is the relation between the perception of the color of a surface and the perception of its illumination and of its spatial position? How does object identification affect the perception of surface color? What is the relation between the perception of the color of a surface and the perception of the texture and material composition of the surface? Answers to these questions are considered in this book, focusing particularly on the results of experiments on how an observer's sensory and perceptual mechanisms affect his perception of surface color.

The study of surface color has developed historically in connection with the problem of color constancy. Theoretical viewpoints have governed much of the research on surface colors; the first chapter outlines these and also considers the problem of specifying the dimensions of an achromatic surface color. The second chapter describes the range of phenomena and problems that the perception of a surface color poses. It also discusses the problem of distinguishing between modes of color appearance and the attributes of a color. The third chapter reviews the many experiments on the effects of contrast and adaptation on color perception. Electrophysiological and psychophysical studies of the physiological mechanisms that underlie the operation of these sensory processes are not included, for many excellent descriptions of these have been published elsewhere. The sensory processes of contrast and adaptation are considered in this book only with respect to the ways they affect the perception of a surface color. Constancy experiments have been performed in many different experimental settings and have involved different theoretical approaches. Chapter IV presents these experimental results. Chapter V reviews theories that explain color constancy in terms of the sensory processes of contrast and adaptation and theories that explain color constancy in terms of the perceptual mechanisms of an illumination frame of reference and of schemata for the perception of visual surfaces. Cues to the illumination, spatial position, and the characteristic color of a surface also affect surface color perception; Chapters VI and VII examine these phenomena. The relation between surface quality and color perception is an important final consideration in Chapter VIII. Chapter IX is a conclusion and summary.

A gap has developed between the early studies conducted under experimental conditions that resemble ordinary experience, and the more recent studies, conducted in a dark room, of textureless surfaces uniformly illuminated. The early studies argue for the importance of perceptual mechanisms in surface color perception, whereas the dark-room studies stress the role of sensory mechanisms. Katz, Gelb, Kardos, MacLeod, and other early investigators sought to study through experiments the fundamental phenomenological observation that a surface color is perceived in a specific illumination. Partly because of the growing influence of sensory physiology, the recent emphasis is on how neural inhibition and summation affect color perception. This book deals with the problem of surface color from a broad perspective, and it is hoped that this approach will be useful in bridging the gap. I have attempted to present experimental results that have not yet received a unified treatment, and what I hope will emerge from such a review is an understanding of the essential processes that underlie the perception of a surface color.

The reader who has no prior knowledge of the subject matter of this book may not find it easy to read. To make the reader's task easier, I have provided a glossary of terms and internal cross references.

I am indebted to Professors James J. Gibson, Julian E. Hochberg, and Robert B. MacLeod, to whom I owe many intellectual debts. It was Professor MacLeod who first interested me in the problem of surface color perception. I am grateful to Professor Fred Attneave, who read a draft of the manuscript, for his valuable comments. I also want to express my appreciation to my wife, Ruth, for her editorial assistance. Excerpts from Beck (1959, 1961, 1964, 1965, 1969, 1971) have been included with the permission of the publishers of the *American Journal of Psychology* and the *Journal of Experimental Psychology*. For permission to reproduce figures, I am grateful to the following publishers: American Association for the Advancement of Science, American Institute of Physics, John Wiley & Sons, Inc., North Holland Publishing Company, Amsterdam, and University of Illinois Press. I am also grateful to Professor Kanizsa for the original photograph for Plate 3, and to Eastman Kodak Company for the original photographs for Plates 5, 8,

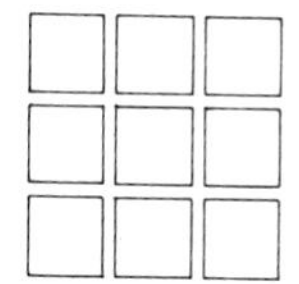

Contents

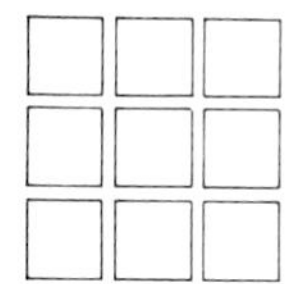

Illustrations

PLATES

FIGURES

TABLES

Surface Color
Perception

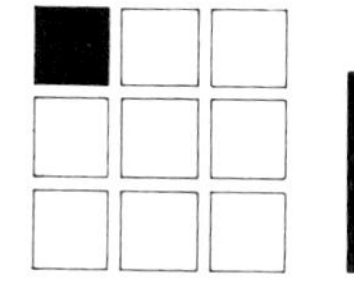

1

Color Constancy

Color constancy refers to the fact that the perceived color of a surface tends to remain constant despite changes in illumination that alter the intensity and spectral composition of the light reflected to the eyes. During the last hundred years a large and varied literature has developed on the many aspects of this problem. The processes that underlie color constancy have not been clearly settled and fundamental theoretical generalizations are not possible. Consequently, one can only present generalized concepts that serve to sort problems into categories and provide a framework for discussion.

The conflicting theoretical viewpoints on constancy may be distinguished by responses to two major questions: Does the perception of a surface color depend on the registering of stimulus information concerning the color of the illumination? Do hypothetical mediators such as frames of reference, schemata, or other perceptual mechanisms that generate descriptions of a visual stimulus intervene between the light reflected to the eyes and the perception of a surface color? Those viewpoints that adopt a negative position to these questions may be described as *sensory theories* and those that adopt a positive position as *cognitive theories*. A compromise position is also possible. Gibson (1966) proposed what might be described as a stimulus theory in which he advocates an affirmative answer to the first question and a negative answer to the second question. Gibson maintains that the perception of a surface color depends upon jointly registering the color of the surface and the color of the illumination but that these are not mediated by perceptual mechanisms. He argues instead that an observer responds to gradients and ratios of stimulation and that these directly specify the color of a surface and the color of the illumination (see p. 158). Gibson has developed his theory primarily in connection with space and form perception and the theory has yet to receive careful formulation for color perception.

Sensory Theories

Sensory theories propose that color constancy is not dependent on stimulus information concerning the color of the illumination. Representative of this viewpoint are attempts to establish a correspondence between sensory processes and the perception of surface color. Cornsweet (1970), for example, proposed that color constancy is the result of simultaneous contrast effects in the visual system that make the perception of color depend upon the relations of intensities within the overall image. Historically, this position goes back to Hering (*1874*, 1964) who, though recognizing the influence of cognitive factors, argued that the major factors in color constancy are the processes of simultaneous contrast and sensory adaptation. Helson (1938, 1943) has also been a vigorous proponent of the sensory point of view. Helson did not relate surface color perception to specific visual processes but defined an intervening variable, the adaptation level, which permits the integration of separate sensory effects. The adaptation level is a function of the light intensities in a scene. When the illumination is changed, the adaptation level changes as a function of the changes in the intensity and spectral composition of the light reflected from different objects in the scene. Sensory processes which alter the adaptation level in proportion to the change in the intensity and quality of the illumination are responsible for the constancy of colors. The contrast and adaptation level theories are considered in Chapter V.

The important factor characterizing what I have called "sensory theories" is obviously not the locus of an effect: contrast and adaptation effects may occur throughout the visual system. Nor are such specific processes as simultaneous contrast and light-dark adaptation important factors. As indicated, Helson's adaptation level principle may be interpreted as an intervening variable permitting the integration of many separate peripheral and central effects. One can argue with some justification that the adaptation level reflects a central transformation process related to the average chromaticity and luminance of a scene as well as local simultaneous and successive contrast (see p. 55 and pp. 80–86). What characterizes the sensory point of view is its insistence that the perception of a surface color and in particular color constancy are to be explained without taking into account the informational or cue properties of stimuli. Surface color perception is to be accounted for in terms of changes in the sensitivities of color response mechanisms due to the complex interplay of summation and inhibition and is not affected by perceptual mechanisms which take into account information about the conditions of illumination or object characteristics. It is invalid, moreover, to assume a difference in the operation of the perceptual processes when a color is perceived in the film and in the surface mode. The impression of illumination that distinguishes the surface mode from the film mode is considered an epiphenomenon. Sensory theories deny that the distinction between surface color and illumination color is a fundamental aspect of surface color perception. Observations that show that an individual's color perception may be affected by how he perceives a surface to be illuminated are presumed to be the result of uncontrolled stimulus variations that affect the sensory interactions. The adequacy of such an interpretation will be considered in Chapter VII.

Cognitive Theories

Cognitive theories have been concerned with accounting for the apparent dependence of perceived surface color upon cues to the illumination and upon the attitude of the observer. The theory of Helmholtz is perhaps the best known of the cognitive theories. Helmholtz (*1867*, 1925) proposed that the perception of color becomes determined relative to an impression of illumination that is derived from the cues provided by the light reflected from the surfaces in a scene. He wrote: "What is constant in the colour of an object is not the brightness and colour of the light which it reflects, but the relation between the intensity of the different coloured constituents of this light, on the one hand, and that of the corresponding constituents of the light which illuminates it on the other. This proportion alone is the expression of a constant property of the object in question (*1868*, 1962, p. 144)." The impression of illumination serves as a reference level for determining the colors of the surfaces in the field. The perception of the achromatic colors white, gray, and black is determined with respect to the intensity of the illumination that the observer registers. White surfaces are those that reflect a great deal of light relative to the registered illumination intensity, and black surfaces are those that reflect

generally held by individuals associated with or strongly influenced by Gestalt psychology, such as Katz (*1911*, 1935), Gelb (1929), and Koffka (1935). Katz and Koffka discarded Helmholtz's assumption that the underlying process is an unconscious inference but retained the hypothesis that surface colors are determined by their relation to a reference level determined by the registered illumination. According to the view of Koffka (1935, p. 171), the perceptions of surface color and illumination color are mutually dependent percepts arising from an organizational process that follows the *Law of Prägnanz,* i.e., the psychological organization will tend toward a "best" structure. If two surfaces are of differing luminance, they will appear to differ in their lightness or in their illumination, depending on the ratios of the light intensities reflected from the surfaces and on the spatial relationships of the surfaces in the scene (see Chapter VII). Color constancy is the consequence of a self-regulating perceptual process that compensates for changes in the quality of the illumination; the process moves toward an equilibrium, or a "reference level of illumination," that results in stable color perceptions. It should be noted that though Helmholtz often spoke of a general inferential process without regard to specific mechanisms, he was well aware that perceptions may depend on specific mechanisms. In volume II of his treatise *Physiological Optics* (*1867*, 1925, p. 274), Helmholtz suggested that the discounting of the illumination occurs in terms of a specific process. The average color reflected from a scene defines an achromatic point. The discounting of the illumination is to be interpreted in terms of a shifting of the achromatic point in the color space (see p. 83). Thus, it is not knowledge of the illumination per se that is important; what is important is the specific information given to the visual system, which determines the achromatic point.

The theory of Judd (1940, 1949) represents a comprehensive and systematic development of Helmholtz's position, taking into account recent studies (see Chapter V).

For both Helmholtz and Koffka the perception of a surface color varies as a function of stimulus variables in similar ways. Since they did not make explicit how stimulus variables determine the registration of illumination, their construct of an illumination frame of reference is not easily distinguishable from the adaptation level principle of Helson. For Helson the reference level established is a function of the luminances of the reflected light weighted for proximity in space and time; for Helmholtz and Koffka the reference level is a function also of the informational properties of stimuli that enable an observer to differentiate between surface color and illumination color. Koffka (1935), basing his views on experiments of Gelb (1930) and Kardos (1934), attempted to explain in terms of a "best" structure why a difference in the composition and the intensity of the reflected light is sometimes seen as a difference in surface color and why it is sometimes seen as a difference in illumination color.

An alternative cognitive theory, alluded to by many, was crystallized by Bartlett (1932) and Woodworth (1938), who applied the idea of schema formation to perception. A schema represents an internal representation of a stimulus that has been built up through a classification of experiences. Bartlett proposed that perception involves the matching of sensory signals to a schema. Woodworth added the important observation that when the perceptual response identifies a stimulus with a schema, note is also taken of the way in which they differ. Beck (1965) has proposed that the perception of a surface color involves two separable components: the signals resulting from the operation of the sensory processes, and the integrative schemata for the perception of visual surfaces into which the signals are assimilated. The sensory interactions of contrast and adaptation provide a basis for the utilization of perceptual schemata which produce a description of variations in the light stimulus in terms of differences in surface color and surface illumination. If surfaces are perceived to be uniformly illuminated, Beck (1965) proposed that when the effects of memory color are ruled out, variations of luminance are perceived as differences in surface lightness, not illumination; this enables lightness perception to be expressed as a function of sensory processes. If surfaces are perceived to be nonuniformly illuminated, however, processes of perceptual organization come into play, for the cue properties of stimuli can cause the observer to see an area of altered surface luminance as an area of altered illumination or as a difference in surface color (see Chapter V). Unlike Helmholtz's and Koffka's hypothesis of an illuminamion frame of reference, the hypothesis that the mediating mechanisms for the perception of a surface color are schemata for the perception of visual surfaces does not imply a precise covariation between the perceptions of illumination and of surface color. The schema hypothesis implies only that cues to the illumination may have an affect on perceived lightness. The conditions under which this occurs will be discussed in Chapter VI.

Dimensions of an Achromatic Surface Color

It is not easy to accurately describe the way things look. The identification of attributes and the treating of them as dimensions raises many difficulties. The specification of the dimensions of a surface color poses a fundamental problem. It is obvious that new dimensions such as from glossy to matte and from fluorescent to nonfluorescent emerge when color is perceived as an attribute of a surface. However, a subtle and difficult problem arises in connection with the dimensions of an achromatic surface color.

One can specify achromatic light by the single psychophysical variable of luminance. The luminance of a sample of light is the radiant energy of that sample weighted in accordance with the photopic spectral sensitivity of the eye. When an achromatic color is perceived in the film or illuminant mode, a single well-defined psychological dimension corresponds to luminance. Luminance variations from just visible to the limit that the eye can tolerate cause achromatic colors to vary along a dimension of brightness, i.e., the variations allow these colors to be ordered on a scale from very dim to dazzling bright.

The attributes that correspond to variations of luminance when an achromatic color is perceived in a surface mode are less clear. The perception of a surface color requires nonuniform stimulation. The simplest condition for the appearance of a surface color is the presence of two neighboring areas of different luminance. This condition is met by a dark disk and light surround. When the luminance of an inlying disk is increased from zero to that of the surround, the perceived color of the disk varies along a dimension of lightness, i.e., allows colors to be ordered on a scale from black through gray to white. The convention adopted by the Optical Society of America (Committee on Colorimetry, 1963) is that there is only a single dimension corresponding to changes in luminance. The dimension that corresponds to luminance is brightness when color is perceived to belong to a film or self-luminous source and lightness when color is perceived as an attribute of a surface.

The phenomenological description of an achromatic surface color is, however, not exhausted by specifying the dimension of lightness. Phenomenological observation suggests that achromatic surface colors may be multidimensional. The two sides of a bent white card, one side illuminated and the other in shadow, are seen as about the same white. The two sides, however, do not appear identical. The unshaded white appears as a brighter or a better white while the shaded white appears as a dull or a poor white. The question arises as to whether the brightness and goodness of a white are true dimensions of a surface color or merely other ways of judging lightness.

Hering (*1874*, 1964) recognized the attributive duality of lightness and brightness. He distinguished between the two attributes of quality (a white-to-black dimension) and weight (a bright-to-dim dimension) of an achromatic surface color. Hering wrote that "whereas the quality of color depends on the ratio of the components, the energy with which it forces itself upon our consciousness, in brief, the expressiveness (*Aufdringlichkeit*) or impressiveness (*Eindringlichkeit*) is determined by its weight (p. 99)." [1] For Hering, lightness and brightness are independent functions of an underlying black-white excitation process. Quality or lightness is a function of the ratio of the white and black components regardless of their absolute magnitudes, i.e., W/(W+BK). Weight or brightness is a function of the absolute magnitudes of the white and black components when the ratios of the components are kept constant, i.e., W+BK.

Katz (*1911*, 1935) proposed that achromatic surface colors are tridimensional. According to him, these colors may vary in insistence (*Eindringlichkeit*) and pronouncedness (*Ausge-*

[1] The term "*Eindringlichkeit*" is more commonly translated as insistence; this is the term I will use. When referring to an achromatic color, the term "brightness" will be used synonymously with "insistence" (see pp. 15 and 28).

prägtheit) as well as lightness. Insistence was defined by Katz as the strength with which a color compels attention and pronouncedness as the degree of goodness of a color—the whiteness of white or blackness of black. For achromatic colors, Katz followed Hering in proposing that insistence varies with the luminance of a surface. Pronouncedness, on the other hand, is a more complex function. Katz claimed that pronouncedness of a white surface increases with illumination while pronouncedness of a black surface decreases with illumination. There has been great uncertainty about the dimension of pronouncedness, and we shall return to a further consideration of pronouncedness later (see p. 106).

For the common experimental setting in which a surface stands in front of a background of higher luminance, the hypothesis of Hering that achromatic surface colors are bidimensional may be formulated in psychophysical terms: brightness is a function of the luminance reflected from a surface (absolute luminance), and lightness is a function of the surface and background luminances (relative luminance). The general view of simultaneous contrast is that the higher luminance of a background decreases the neural intensity signal of an inlying disk. A quantity determined by the background luminance is assumed to be subtracted from the neural correlate of the luminance of the disk because of lateral inhibitory interactions. Information about the absolute luminance of the disk is presumably not transmitted by the altered neural output. Rather, the neural signal resulting from lateral inhibition provides information for the perception of the lightness of the disk. It is possible, however, that the sensory system preserves information about both the unaltered (absolute luminance) and altered (relative luminance) neural signals, i.e., provides for separate luminance detectors and contrast detectors. Wallach (1963) and Evans (1964) assumed that information about both absolute luminance and relative luminance is transmitted by the nervous system. Wallach (1963) assumed that light stimuli produce two effects: a bright-to-dim color impression that is dependent on the absolute luminance of a stimulus and on the adaptation of the eye, and a white-to-black color impression that is dependent on the interaction between neighboring regions of differing luminance. Evans (1964) proposed that whenever a luminance is lower than another, it takes on a gray appearance. The variable of grayness or gray content is independent of the variable of brightness, which is a function of the absolute stimulus luminance.

An alternative view is that information about the absolute luminance of a surface is not preserved under the contrast interaction. Information about the absolute luminance of a surface is suppressed when lateral inhibitory interactions alter the neural signal. The observer is not able under these conditions to assess luminance by judging the brightness of a surface. The distinction between the dimensions of brightness and lightness of an achromatic surface emerges rather at a central level. Brightness is not the perception of the absolute luminance of a surface but the perception of the intensity of the illumination of a surface. Katz (*1911*, 1935) showed that an individual can become aware of the intensity of the illumination of a surface. Awareness of the illumination is made possible by the reflected luminances in a field. There are many variables of the reflected luminance that can provide information about the intensity of the illumination. Beck (1959, 1961) found that illumination judgments of a surface in the experimental conditions that he investigated were strongly influenced by the luminance of highlights, or clearly discriminable white areas of a surface or a background (see p. 100). The neural intensity signals of highlights and of areas seen as white are not modified by lateral inhibition and vary directly with the intensity of the illumination. What is suggested is that the perception of the brightness of an achromatic surface color is a function of surface highlights and other properties of the reflected light that indicate the intensity of the illumination of a surface. A bright black surface is a black surface perceived to be strongly illuminated; a dim white surface is a white surface perceived to be weakly illuminated. Boring (1942), Koffka (1935), and Thouless (1932) appear to have identified the attribute of brightness of an achromatic surface color with the perception of the intensity of the illumination. The evidence for the two alternative hypotheses offered will be summarized in Chapter VI.

Questions analogous to that raised about whether the process of lateral inhibition suppresses information concerning the absolute luminance of a stimulus may be raised for each stage of the processing of intensity information in the visual system. A physical stimulus produces a complex sequence of processing stages. For example, three stages that may be distinguished are: the primary stimulation of the rods and cones; the modified output resulting from contrast interactions; and the output from the operation of interpretative mechanisms such as schemata. At each stage of processing, information about the intensive properties of a stimulus encoded at previous stages may be suppressed or transmitted. There is, therefore, no a priori limit to the number of possible intensive properties of a stimulus. The attributes of a stimulus depend on how the information is transformed at each stage of processing. What information is suppressed and what information is transmitted?

Beck (1965) has proposed that the output of sensory processes does not specify a particular color percept but allows for alternative color percepts. Changes in the perception of color may be brought about by assimilating sensory signals to a different perceptual schema. For example, a stimulus may be seen as a gray or as a white in reduced illumination, depending on the attitude or set of the observer (see p. 66). The question may be raised whether there does not remain something constant with such shifts of attitude. Woodworth and Schlosberg (1954) suggested that the combination of the perceived lightness of a surface and the perceived intensity of the illumination is phenomenally constant with shifts of attitude. It is possible that changes in the "interpretation" of a stimulus due to shifts in observational attitude need not alter the sensory signals. An individual who observes with a sophisticated and critical attitude can become conscious of the information corresponding to the neural intensity signal resulting from the sensory processes.

The trouble with the word "brightness" is that it has been used in a great many different senses in connection with an achromatic surface color. It is often used to refer to the perception of what we have called lightness. "Brightness" has been used to refer to the quality Hering and Katz called *Eindringlichkeit*. "Brightness" has also been used to refer to the perception of the absolute luminance of a surface as in the hypotheses of Hering and Katz, and it has been used to refer to a hypothetical absolute intensive magnitude of a color that does not change as a function of the "interpretation" of a stimulus (Bartelson and Breneman, 1967). The "brightness" impression is assumed to remain constant though the perceived lightness of a surface may change (e.g., from gray to a white dimly illuminated) because of apperceptive processes. The word "brightness" has in addition been used to refer to the perceived intensity of the illumination (Koffka, 1935). Clarification of terminology is certainly in order.

I shall use the term *brightness* to refer to the quality of an achromatic surface color which Hering and Katz called *Eindringlichkeit* (insistence) (pp. 10–11). I shall use the word *lightness* to refer to the white-to-black dimension of a surface color. The word *brightness* will also be used to refer to the bright-to-dim dimension of a film or self-luminous color. I shall not use the word "brightness" in referring to the apparent illumination but shall speak instead of the *perceived intensity of the illumination*. The context will make clear when "brightness" is used to refer to the perception of the absolute luminance of a surface as in the hypotheses of Hering and Katz. The phrase *apparent luminance* will be used to refer to another hypothesized percept—the intensive magnitude of a surface color that is presumed to remain constant with changes in lightness due to the interpretation of a stimulus situation.

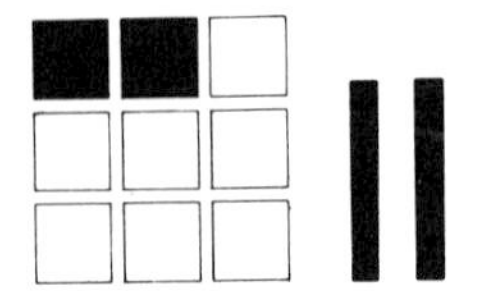

Modes of Color Appearance

In 1765, Thomas Reid pointed out that perception is normally of objects and not of sensory qualities. The implications of this observation for color perception were not systematically considered until Katz's (*1911*, 1935) classic work, *The World of Color*. Katz was the first to attempt a detailed presentation of the complex ways in which colors are perceptually experienced. He called these "modes of color appearance." The phenomenological approach adopted by Katz was to describe as accurately as possible the phenomena of color in different modes of appearance. He described and compared eleven modes of appearance; the more important are film colors (*Flächenfarben*), surface colors (*Oberflächenfarben*), volume colors (*Raumfarben*), mirror colors (*gespiegelte Farben*), luminous colors (*Leuchtende Farben*), the illumination of empty space (*Erleuchtung*), and illumination of an object (*Beleuchtung*). Plate 1 shows the appearance of colors in different modes.

Evans (1948, p. 168) has suggested that Katz's eleven modes may be included under three general modes of appearance that classify color according to the three fundamental ways in which it may be apprehended: as a film without

Plate 1. The different modes of appearance of color: film, illuminant, volume, and various surface colors.

objective reference, as a property of an object, and as a property of illumination. The classification developed by the Committee on Colorimetry of the Optical Society (1953) is similar. Table 1, adapted from *The Science of Color,* lists five modes of appearance and the attributes of each. The surface and volume modes are subtypes of the object mode; the illumination of empty space and the illuminant modes are subtypes of the illumination mode. It is convenient to use a small number of general modes for a taxonomic description of color appearances. The more detailed studies of Katz are valuable for a phenomenological description as a propaedeutic undertaking, which ensures that concepts and principles proposed are adequate to account for all problems raised by the perception of color.

I shall begin by briefly describing important instances of colors in the film, object, and illumination modes. I shall then consider the meaning of the concept of modes of appearance. Following this I shall describe the attributive and functional properties of film and surface colors.

Film, Object, and Illumination Modes

Film Colors. A film mode of appearance occurs when stimulation has been reduced in such a way as to eliminate all information about the material composition and spatial location of an object. Film colors always appear as an expanse of color extended in space in the form of a bidimensional plane —as in a spectroscope. A film color is perceived when an observer views a far surface through a small hole in a screen placed at some distance from the eyes. Because the observer is unable to accommodate for the texture of the far surface, an area of uniform luminance and chromaticity is projected on the retina. The area within the hole has a hue, brightness, and saturation, but is no longer perceived as a material

Table 1. The attributes and dimensions of perceived colors classified according to modes of appearance.

Attributes and dimensions	Modes of appearance				
	Illuminant (glow)	Illumination (fills space)	Surface (plane object)	Volume (tridimensional object)	Film (aperture)
1. Hue	x	x	x	x	x
2. Saturation	x	x	x	x	x
3. Brightness	x	x	d	d	x
4. Lightness			x	x	
5. Duration	x	x	x	x	x
6. Size	x	(x)	x	x	(x)
7. Shape	x	(x)	x	x	(x)
8. Location	x	(x)	x	x	not in depth
9. Texture	a		x	x	
10. Glossiness (lustre)			x		
11. Transparency	(x)	x		x	
12. Fluctuation (flicker, sparkle, glitter)	x	x	x		
13. Insistence	b	b	b	b	b
14. Pronouncedness	c	c	x	x	c
15. Fluorescence			x	x	

Adapted from *The Science of Color,* Committee on Colorimetry, Optical Society of America (New York: Thomas Y. Crowell Co., 1953), p. 151.

( ) Parentheses indicate that in the writer's opinion the attribute occurs in a limited or indefinite manner.

[a] Texture tends to produce a surface appearance rather than an appearance of a self-luminous source. At high luminances a textured surface may look like a source.

[b] For achromatic colors, insistence is equal to brightness (p. 28).

[c] According to the interpretation proposed on pp. 106–108, pronouncedness in the illuminant, illumination and film modes occur only for chromatic colors.

[d] It should be noted that in the convention adopted by the Committee on Colorimetry (1953) there is only a single dimension corresponding to luminance—brightness when the color is perceived to belong to a film or self-luminous mode and lightness when the color is perceived to belong to an object mode (see pp. 9–13).

surface, as illuminated, or as having a definite spatial location except at an indeterminate distance behind the screen. The screen is typically of lower luminance than that of the surface behind the hole. Optically, the stimulus for a film color consists of an area of uniform luminance and chromaticity separated by a sharp contour from an area of lower luminance.

Object Colors. The most common occurrence of color in the object mode is as an attribute of a surface. Katz (*1911*, 1935) described a surface as: (a) being visually "hard," i.e., appearing to be a material object; (b) having a definite color; (c) standing in a definite illumination; and (d) having a definite spatial location. Gibson (1950) listed eight phenomenal qualities of a surface: (a) visual resistance, or "hardness"; (b) color; (c) illumination; (d) slant; (e) distance; (f) contour; (g) shape; and (h) size. Not all these properties are necessary for the perception of a surface. Gibson points out that if an observer stands close to a wall, the surface that he perceives has no determinate contour, shape, or size. However, he does perceive a material surface of a definite color, illumination, and location. Whether all three of these properties are necessary for the perception of surface color is not definitely known. The experiments by Kardos (1929), Wallach (1948), and Beck (1959) suggested that stimuli complex enough to give rise to the perception of a surface contain information for the perception of the color of the surface and the color of the illumination. Martin (1922) reported that under the conditions of her experiment the localization of a color at a determinable distance is a necessary and sufficient condition for the shift from a film to a surface appearance. Under these conditions of observation a stimulus may be equivocal, and an observer, depending upon his attitude, may see either an indefinitely localized film color or a surface color located in a given position. Under ordinary conditions of observation the perception of a surface color is determined by the stimulus and not affected by an observer's perceptual attitude. The differential reflection of light by the physical and chemical inhomogeneities present in a surface is the stimulus for the perception of that surface. Two areas of different uniform intensities are sufficient to give rise to a surface appearance when they are contiguous and separated by a sharp contour (Wallach, 1948). If the two areas are separated by a gap, they will not be perceived as surface colors but as self-luminous lights of differing brightness.

Object colors can also occur in relation to the perception of tridimensionality. Volume colors refer to colors seen to fill a definite space, e.g., a transparent surface such as a red color filter. The redness of a red glass or a red liquid is perceived both in and behind the surface. Volume colors are at least partially transparent, and objects may be seen through them. In the limiting case volume colors are seen as transparent films, as, for example, light falling on the blades of a rapidly rotating fan. Related to the transparency that is seen occurring in volume colors is the distinctive integration of color and depth that is seen to occur in mirror colors. Mirror colors are an example of what Koffka (1935) called duo-organizations. One color is seen behind another color. In looking at the reflection of a color in a polished surface, the color of the object is seen behind and through the color of the reflecting surface. A common example is the reflection of a white surface in a black, glossy tile. The appearance of one color behind another always seems to involve a difference in depth.

Illumination Colors. Color in the illumination mode is perceived as a property of illumination. One example is illuminant color. Illuminant colors, as in the case of lamps, may have a definite location and object character. However, they are seen as a source of light with the light coming from behind the surface rather than as illuminated objects.

Katz (*1911*, 1935) must be credited with calling attention to the perception of the illumination of empty space as distinct from the perception of objects and surfaces. In this mode, colors are apprehended as an attribute of the space between objects rather than as an attribute of the objects. There has been considerable controversy about the way in which the lighting of a spatial region is apprehended. Koffka (1932), for example, suggested that the intensity of the lighting of space is not really "seen" but is rather "felt" by an observer. Similarly, Woodworth (1938) spoke of "registering" the illumination to indicate that the illumination present may be taken account of without being part of conscious experience. We ordinarily center our attention on the objects before us and not on the illumination and therefore often have no pronounced impression of the illumination. When one turns his attention to the illumination, however, it seems to be a fact that one can become aware of the lighting prevailing in a region. Looking into a neighboring room one has a definite impression of the intensity and color of the light in the room. This impression of illumination is particularly evident when the lighting in the neighboring room differs from the lighting in the room in which one is located.

The illumination of an individual object or surface can also be perceived. One way in which we see illumination on a surface is in terms of cast shadows and light spots. Light spots refer to the diffuse light that appears as excess light on a surface. Both shadows and light spots are phenomenologically distinct. In fact, they may at times be seen as independent films separable from the surface on which they lie

(MacLeod, 1932). We are also able to perceive variations in the intensity of the illumination of a surface; for example we notice how the illumination of an object changes when the object is moved toward or away from a light source, although we do not perceive the illumination in this case as an independent entity separable from the surface in the way that we often see shadows and light spots (see Chapter VI). We are also able to perceive changes in the color of the illumination.

Concept of Modes of Appearance

What is the significance of Katz's concept of modes of appearance? Is there a difference between the concept of mode of appearance and the concept of an attribute? Troland (1929, p. 107) distinguished between a description of experience in terms of attributes and in terms of structural relations. A description in terms of attributes involves abstracting the ways in which aspects of a percept can vary. A structural description describes the relations between separable components of the percept. Though there has been some vagueness in past usage, the concept of modes of appearance may be interpreted to refer to the structural relations that occur in the perception of color. Film, object, and illumination colors represent general categories that are associated with distinctive integrations of properties that elicit a nonreferential, object, and illumination color experience. Subcategories of different modes of appearance describe perceptual structures of varying relationships. For example, in the object mode, the perceptions of an illuminated surface, of one surface behind another (as in mirror colors), and of two sides of a cube, one side of which is seen facing the light and one side of which is seen turned away from the light, all involve particular structural relations between the perceptions of surface, space, and illumination. According to the interpretation I propose above, modes of appearance refer to perceptual categories like apples and oranges rather than to attributes. The probability that sensory signals will be categorized in terms of a film, surface, or illuminant mode is not only a matter of the fit between the sensory signals and the category specification. It may depend also on the attitude of the observer. As mentioned before, alternative modal perceptions are possible in the case of equivocal stimuli, depending on the attitude of the observer. What alternative modal appearance can occur is limited by the stimulus information. Under ordinary conditions it is impossible for the observer by change of attitude alone to see cloth or wood as anything but a surface color.

Modes of appearance are not isolated categories; they may also exhibit transitional appearances. However, these transitional appearances are not true phenomenological intermediates in the sense of the intermediate qualities that lie between two values of an attribute, as, for example, orange is a distinct hue that lies between red and yellow. Martin (1922) reported that film and surface colors are separated by a qualitative gap. Intermediate appearances exist only in the sense of equivocal stimuli that can be seen either as a film or as a surface depending upon an observer's attitude, or of stimuli in which parts of the stimulus look like a film and parts of the stimulus look like a surface. Gibson, Purdy and Lawrence (1955) reported that a surface with less than twenty contours appears filmy and insubstantial. Observers' descriptions indicated that, instead of a continuous surface, they perceived pieces of surface with gaps between them. That is, the percept combined both surface and film appearances rather than being a true intermediate between the surface and film modes. Martin (1922) also found no true phenomenological intermediates between film and volume colors. Avant (1965) has reviewed the research on the transitions between the perception of a surface and the experience of a mist of light that occurs with complete homogeneous retinal stimulation (a *Ganzfeld*). Though it is difficult to be certain, there also seem to be no true intermediates between the appearance of a surface and the appearance of a *Ganzfeld*.[2]

Modes of appearance and the attributive and functional properties of color are interrelated. Certain attributes such as hue, saturation, and brightness are present in all modes of appearance. An investigation of these attributes in the film mode has been useful for investigating the sensory processes in color perception. It is, however, an abstraction to consider the perception of color apart from modes of appearance. A naïve observer, for example, finds it difficult to make color comparisons when modes of appearance differ, as when he is asked to match the color of a cloth with the color of a vial of liquid or with the color of a lamp. Many attributes occur only in connection with certain modes of appearance. The color appearance "clear" occurs in modes of appearance in which transparency is possible. It can, therefore, occur in the object and illumination color modes but not in the film color mode. For example, in connection with the illumination of space, the interspaces and spatial relations of a dimly lighted room are the same as those of a well

[2] The perception of fluorescent colors presents an interesting phenomenological problem. Under ordinary observing conditions, a fluorescent color appears as a surface color of unusual brightness and saturation. In a dark room, a fluorescent color may appear either as a surface color or as a luminous color. When seen as a luminous color, however, the light appears to come from within the surface. Fluorescent colors may, therefore, be ambiguous in that they may appear either as a surface or an illuminant color, but they do not represent an intermediate mode between the surface and illuminant modes.

do not respond to colors only in terms of these three dimensions. A film color of high brightness and saturation may be described as brilliant; a film color of low brightness and saturation as dull. Stevens (1934) suggested a more liberal criterion for defining a psychological dimension, that of independent constancy: a dimension must be capable of being held constant while other dimensions are varied. A criterion of independent constancy adds phenomenologically distinct, but nonindependent, dimensions. The perceived chromaticity of a color, i.e., the chromatic power of a color, is a function of the dimensions of hue and saturation and meets the criterion of independent constancy. It is possible to plot loci of constant perceived chromaticity while hue, saturation, and brightness vary. An important problem in colorimetry is to determine a spacing of colors that would accurately reflect chromaticity differences (Judd & Wyszecki, 1963). Colors may also differ with respect to affective attributes such as warm and cold, pleasant and unpleasant, and so on. The number of dimensions of film colors satisfying a criterion of independent constancy has never been fully determined.

The perception of surface colors introduces new independent dimensions of color. Surface colors vary in lightness as well as in hue and saturation. Lightness is the dimension in the achromatic color series ranging from white through gray to black. The dimension of lightness emerges when there are differing luminances in neighboring regions. The perception of lightness is not restricted to achromatic colors. Lightness variations also occur in connection with hue. The colors brown, olive-green, and maroon are examples of changes in lightness. Brown is the result of adding dark gray to an orange color, olive-green of adding dark gray to a green color, and maroon of adding dark gray to a red color. These colors differ from those like red, green, and blue. The latter correspond to differences in the wavelength or composition of wavelengths that stimulate the eye, which is not the case with colors that are variations in lightness. The spectral composition of the light reflected by an orange peel and a chocolate bar, for example, can be identical. Only the percentage of the light reflected is different. If the light reflected from an orange peel and a chocolate bar were reconstituted as arrays of light of uniform luminance in the dark, i.e., film colors, the two colors would match in hue and saturation and differ only in brightness; the light from the orange peel would give a brighter color than the light from the chocolate bar. The dimension of lightness usually occurs only with object colors. Film colors may vary from dim to bright, but one does not ordinarily experience a gray film color.

Lightness is not the only new dimension that emerges with surface colors. Evans (1959, p. 149) has suggested two: the glossy-to-matte dimension and the rough-to-smooth dimension. These dimensions of surface color are related to the distribution of intensities of the reflected light as a function of the angle of incidence and of the physical texture of the surface. Glossiness is a complex attribute in the sense that it is a response of the visual system to the spatial and temporal patterning of the light stimulus. For a complex attribute, such as glossiness, many transitional series exist between a perfectly matte and a perfectly glossy surface (Judd & Wyszecki, 1963, p. 373). The rough-smooth dimension has been studied experimentally only in connection with the perception of graininess (Jones & Higgins, 1947; Zweig, 1956). Like the glossy-matte dimension, the rough-smooth dimension is complex; there are undoubtedly many types of transitions between rough and smooth surfaces. The variations in surface texture and finish are many, but a review of the literature by Pickett (1968) indicates that they have been little studied.

Characteristic patterns of reflectance also produce colors associated with specific objects. Thus, silver, gold, and copper are colors that are associated with the particular appearance of metals. These metallic colors result from selective specular reflection from the top surface of the metal. Metallic colors, like glossy colors, do not correspond to a single variable of physical energy, but to a complex pattern of reflectance. One may determine the hue of the specularly reflected light, but it will not have the characteristic appearance associated with the metallic color; e.g., gold strongly reflects red and yellow light (Kodak, 1950). Though the conditions for the appearance of metallic colors are well known, it is uncertain whether the range from metallic to nonmetallic is a dimension. Another color closely associated with specific objects is the color impression of clear that occurs with transparent surfaces such as panes of glass. Evans (1959) points out that a white-to-clear dimension is associated with the degree to which a surface reflects light diffusely as compared to transmitting it regularly. Temporal patterns of stimulation produce still other color attributes such as sparkle and flicker.

Katz (*1911*, 1935) has also distinguished two dimensions of a surface that vary with the reflectance and with the illumination of the surfaces: insistence and pronouncedness. According to Katz, both achromatic and chromatic surface colors may vary in either insistence or pronouncedness while maintaining constant the attributes of hue, saturation, and lightness. For chromatic colors, insistence and pronouncedness are functions of the attributes of hue, saturation, and brightness. For achromatic colors, Katz (*1911*, 1935) proposed that insistence is a direct function of the absolute luminance and is determined by the reflectance and illumination of a surface. Thus, the insistence of an achromatic surface increases with increasing illumination and corresponds to the perception of the absolute luminance of a surface. The bright-to-dim dimension of a color is most clear in the perception of self-luminous colors. Brightness judgments of self-luminous sources are readily made. Observers vary considerably in their ability to make judgments of the brightness of surface colors, and matching two surfaces for equal brightness may be very difficult for some observers (see p. 104). Recently, considerable interest has been aroused in fluorescence, a dimension of surface colors that meets Steven's criterion of independent constancy but not Titchener's criterion of independent variability. When the hue is constant, the appearance of fluorescence varies as a function of the lightness and saturation of a surface color (see Chapter VIII).

Functional Properties of Film and Surface Colors

The central problem of surface color perception is that of color constancy. As previously indicated, this phrase refers to the tendency of colors to retain approximately the same appearance despite wide variations in the luminance and spectral composition of the light stimulating the eyes. The intensity and spectral composition of the light reflected from a surface are a joint function of the incident illumination and the reflectance of the surface. Whenever the intensity or spectral composition of the illumination is altered, the light reflected by the surface is automatically altered. However, the perceived hue, saturation, and lightness of a surface retain their daylight appearance despite considerable changes both in intensity and spectral composition. In fact, the perceived color of a surface is often better predicted by the reflectance characteristics of a surface than by the intensity and spectral distribution of the reflected light.

Katz (*1911*, 1935) asserted that film colors are in correspondence with the retinal excitation and do not, therefore, exhibit constancy. Surface colors, in contrast, exhibit color constancy. This generalization requires qualification. The reduction screen experiment previously described (see p. 17) furnishes a paradigm whereby all colors can be reduced to the film mode of appearance. Under these conditions, there is an elimination of contrast effects with neighboring areas in the visual field and an elimination of the properties of stimulation that specify the spatial location, shape, slant, surface texture, and amount and direction of the illumination of the surface. If the adaptation of the eyes is taken into account, the color perceived as belonging to a film seen through a hole in a reduction screen is in correspondence with the fundamental physiological responses of the retinal color receptors. In this sense Katz was correct in asserting that a film color is in direct correspondence with the degree of excitation of the retinal receptors. One should note, however, that a film color seen through a reduction screen will exhibit constancy with changes of illumination to the extent that adaptation is effective in maintaining the appearance of a constant color.

Maximum color constancy occurs with free binocular observation of surface colors under ordinary conditions. Under ordinary conditions, the effects of contrast and adaptation, the stimulus information for the spatial arrangement of surfaces, the amount and direction of illumination, and memory color may all operate to maintain constancy. Though phenomenologically there appears to be a sharp distinction between the film and surface modes, no dichotomy exists between film and surface colors with respect to the question of constancy. A surface color may exhibit little constancy under certain conditions of observation (see p. 70). More generally, reduction can be thought of as a process of eliminating those properties of stimulation that serve to maintain constancy. A film color observed in a dark room represents complete reduction. There are, however, intermediate stages of reduction. The effects of memory color may be eliminated by using a target that has no familiar color associated with it. Further reduction may be effected by concealing the spatial relationships between a target and neighboring surfaces, and by eliminating information about the amount and the direction of illumination. With complete reduction, the effects of sensory contrast are also eliminated. The extent to which constancy fails under varying degrees of reduction provides a basis for inferring the underlying processes involved in the perception of a surface color.

Color constancy cannot be specified in terms of stimulus variables alone. Alternative modes of appearance to the same stimulation are possible depending upon attitude or set that may affect the perceived color. For example, Evans (1948) pointed out that the attitude of the observer is important in the perception of mirror colors. One may adopt two different attitudes in looking at the reflection of a white card in a black tile. If one takes an attitude that localizes the reflection in the same plane as the reflecting surface, a gray is seen. If the card is localized behind the reflecting surface, a white card is seen behind a black tile. As mentioned before, the perception of a color in the surface mode involves perceiving the surface as having a color in a given illumination. Observation of a surface through a reduction screen that destroys the discrimination between the color of the surface and the color of the illuminant eliminates constancy. The separation of illuminant color and surface color may also be affected by an observer's attitude. Kardos (1928) found that under conditions of reduction that produce an equivocal stimulus an observer exhibited greater constancy when set to perceive a surface under changing illumination than when set to perceive a film color. Judd (1960) has proposed that the mode of the perceived color, whether film or surface, is of basic importance in determining the way in which sensory signals are processed. The impression of a surface is fundamental to explanations of constancy in which an important factor is the separation of illuminant color from surface color (Judd, 1960; Beck, 1965; Gibson, 1966). The perception of a surface rather than a film, however, does not in all circumstances improve constancy. Helson (1938) found no differences between the constancy of surface and film colors in his studies of color adaptation. The fact that under certain conditions the mode of appearance does not affect color constancy does not mean that under other conditions mode of appearance will not affect constancy. The basic question is: Under what conditions does mode of appearance affect color perception and under what conditions does it not? (See p. 85).

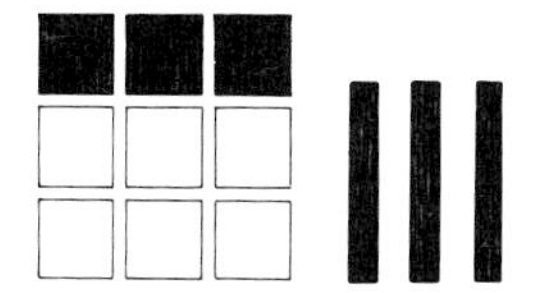

III

Contrast, Assimilation, and Adaptation

Katz's phenomenological study of the modes of color appearance emphasizes the fact that, although one can abstract the sensory qualities of color, the perception of color in everyday experience is intricately related to the perceptions of space, surface, and illumination. The structural relations occurring in surface color perception bring to the fore questions of how the localization of surfaces, their perceived surface qualities, and their perceived illumination determine the seen color. The effects of these factors depend on integrative and organizational processes. Before turning to how these more complex processes may affect color perception it is necessary to review experiments on the sensory aspects of color perception. This chapter considers effects arising from processes of inhibition and summation.

Simultaneous Lightness (Brightness) Contrast [3]

Simultaneous contrast refers to the changes in the lightness or brightness of a stimulus (called a test field, TF) due

[3] A variety of experimental procedures have been employed in the study of contrast. One procedure is to compare the appearance of a test field subject to contrast with a matching, or reference, field so pre-

to the presence of a neighboring stimulus (called an inducing field, IF) of differing luminance. In general, if the luminance of the test field is higher than that of the inducing field, the test field appears self-luminous, and contrast changes the brightness of the test field (bright-to-dim dimension). If the luminance of the test field is below that of the inducing field, the test field appears as a surface color and contrast changes the lightness of the test field (white-to-black dimension). Simultaneous contrast represents essentially an instantaneous adjustment of the visual system and can produce dramatic effects in color perception. The phenomenon of simultaneous contrast makes color perception depend upon

---

sented as to be uninfluenced by contrast. To eliminate possible contrast effects, the matching field is presented in a dark surround to one eye and the test and inducing fields to the other eye. A measure of the contrast produced by the inducing field is obtained by adjusting either the luminance of the matching or of the test field so that the two appear of equal brightness. The matching field varies in the bright-to-dim dimension because of the absence of an inducing field whereas the test field varies in the bright-to-dim dimension when the luminance of the inducing field is below that of the test field and will vary in the white-to-black dimension as the luminance of the inducing field is increased above that of the test field. The term "brightness" is used to refer to the apparent intensity of a test field in both the self-luminous and surface modes. To be consistent with our previous definitions (see p. 15), we will use the term "lightness" to describe the effects of contrast when the test field varies in the white-to-black dimension and "brightness" when the test field varies in the bright-to-dim dimension. It should be remembered, however, that statements about changes in the lightness of a test field in such experiments are inferences from the matching data. The effects of the inducing field on the lightness of a test field can be specified more certainly in experiments such as Hess and Pretori (1894) and Wallach (1948) in which both the test and matching fields were surrounded by inducing fields of higher luminance and varied in lightness.

the relationships between luminances rather than upon their absolute values. Simultaneous contrast varies as a function of the luminances of the test and inducing fields and of their spatial configuration. Two spatial arrangements have been investigated: the inducing field surrounds the test field, and the inducing field is adjacent to the test field.

The first systematic study of simultaneous contrast was by Hess and Pretori (1894). Their results showed that a test field surrounded by an inducing field of higher luminance will appear darker as the luminance of the inducing field increases. The basic relationship found by Hess and Pretori is that the luminance of a test field may be adjusted to maintain a constant lightness for many different luminances of the inducing field. Heinemann (1955) used an annular inducing field 55 minutes in diameter that surrounded a central circular test field 28 minutes in diameter. He found that if an inducing field is of a lower luminance than the test field the brightness of the test field either remains the same or is slightly enhanced above what it would be alone. As the luminance of the inducing field is made greater than that of the test field, the lightness of the test field decreases. Large decreases in lightness apparently occurred when the luminance of the inducing field was only one and a half times that of the test field. Similar results have been reported by Horeman (1963) and Torii and Uemura (1965). Torii and Uemura used a test field 36 minutes in diameter surrounded by annular inducing fields of .8, 1, and 2 degrees. They found that the rate of decrease in the lightness of a test field became steeper with both increases in the luminance and size of the inducing field. Heinemann (1955, 1961) also reported that the perceived lightness of a test field remains approximately constant over a range of inducing field luminances if the ratio of the luminance of the test field to the luminance of the inducing field remains constant. That is, in the equation $TF/IF = TF^*/IF^*$ (where TF and TF^* represent the luminances of two test fields, and IF and IF^* of two inducing fields) the luminance ratios of the test fields to inducing fields are equal, and therefore the test fields are perceived to be of nearly equal lightness. It should be noted that although changes in the intensity of the illumination change the absolute amounts of light reflected by different targets, the ratios of the light intensities reflected remain constant. If Heinemann's finding held generally, constancy of lightness would be almost perfect. But although the finding that the lightness of a test field remains approximately constant when *TF* to *IF* luminance ratios are equal has been reported by a few other investigators (Wallach, 1948; Horeman, 1963), other researchers have had dissimilar findings. Stevens and Stevens (1960) investigated contrast as a function of the absolute intensities of a test field and surrounding inducing field when the luminance ratio of the test field to the inducing field was fixed. The results obtained are shown in Figure 1. As the level of the illumination is raised, the lightness of a light gray becomes lighter, the lightness of a medium dark gray remains the same, and the lightness of a black decreases. Similar results have been obtained by Jameson and Hurvich (1961). Their experimental arrangement consisted of five squares of differing lightness arranged in a cross. As in the experiment of Stevens and Stevens (1960), light squares increased in lightness, a square of intermediate lightness remained constant, and a black square became still darker with increasing illumination. Bartelson and Breneman (1967) reported data on lightness judgments of a complex photographic scene at various levels of illumination that agree well with the Jameson and Hurvich data for the cross arrangement. Hurvich and Jameson (1966) have also

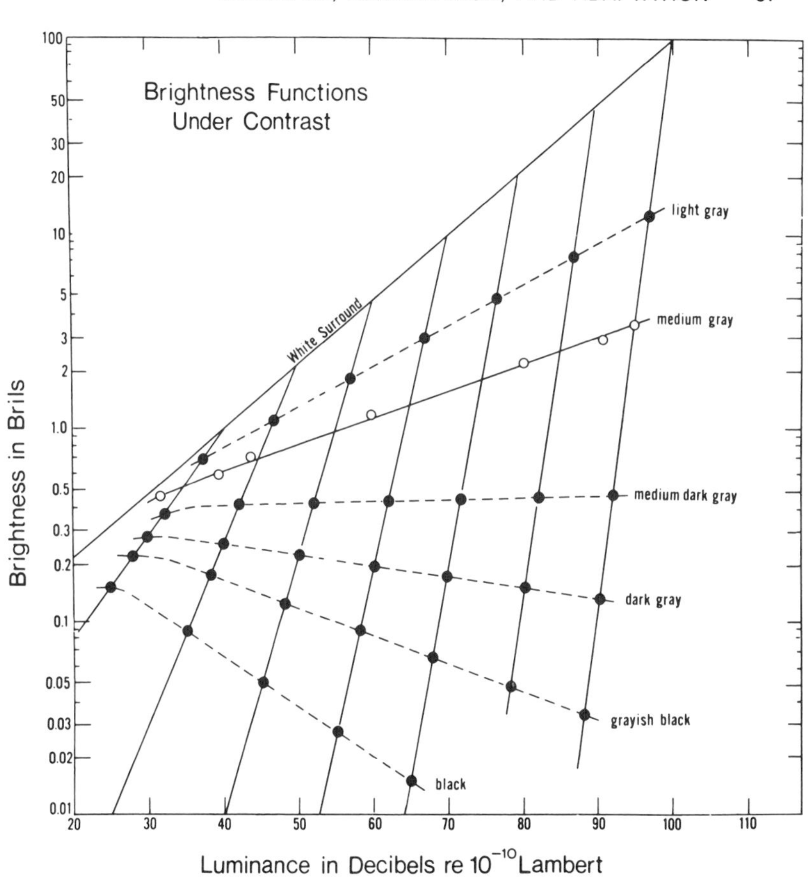

Figure 1. Brightness functions under contrast, showing the appearance of gray papers of differing reflectance viewed against a white background under various levels of illumination. The dashed lines represent fixed ratios of target-to-background luminance. The filled circles plot the data obtained by Leibowitz et al. (1953). The bril scale of brightness describes observers' judgments of the apparent intensity accompanying changes in luminance. The scale applies to both changes in the bright-dim dimension of a luminous source and to the black-white dimension of a surface color. (Adapted from S. S. Stevens, To Honor Fechner and Repeal his Law, *Science*, Vol. 133, 13 January 1961, 85. Copyright 1961 by the American Association for the Advancement of Science.)

reanalyzed the results of Hess and Pretori and have showed that these results follow a similar course.

When the inducing field and test field are adjacent, the overall contrast relationships are similar to those that occur when the inducing field surrounds the test field. Diamond (1953) reported that an adjacent inducing field 33 minutes by 33 minutes increasingly depresses the lightness of an equal-sized test field as its luminance exceeds the luminance of the test field, although the reduction in lightness is much more gradual than that reported by Heinemann (1955). However, Diamond found no evidence for brightness enhancement when the luminance of the inducing field was below that of the test field. Similar results were found by Leibowitz, Mote, and Thurlow (1953) and Horeman (1963). Horeman (1963) replicated the conditions of Heinemann (1955) and Diamond (1953) and compared the contrast effects when the inducing field surrounds the test field and when the inducing field is adjacent to the test field. He found that the magnitude of the contrast effect is greatly reduced when the inducing field is adjacent to the test field rather than surrounding it. It should be noted, however, that a variable that is confounded with spatial arrangement is the relative sizes of the inducing and test fields. Diamond's inducing field was three times larger than the test field when the IF surrounded the TF; the inducing and test fields were of equal size when they were adjacent to each other. Diamond (1955) studied the effect of the size of the inducing field. A rectangle 33 minutes by 16.5 minutes was used as the test field. The inducing field was 33 minutes in length and varied in height from 0 to 33 minutes. As the size of a brighter inducing field was increased, the lightness of the test field was decreased. Diamond (1962) found that similar variations in the size of the test field does not affect perceived lightness.

Leibowitz, Mote, and Thurlow (1953) and Fry and Alpern (1953) studied the effect of the spatial separation between the test field and the inducing field and found that the contrast effect decreases sharply as the distance between inducing and test fields increases. In the Fry and Alpern study the test field was a .5 degree by 2.5 degree rectangle centered between two inducing fields of the same size. At the maximum separation investigated between the test and the inducing fields (4.5 degrees), minimal contrast effect occurred when the luminances of the inducing fields were less than 3,200 foot-Lamberts (ft.-L.). At this separation with inducing field luminances of 10,000 ft.-L. and above there was a considerable contrast effect and darkening of the test field. Leibowitz et al. reported minimal contrast effects when the separation between a test field 30 minutes by 30 minutes and an equal inducing field was greater than 9 degrees. The amount of contrast with spatial separation of the inducing and test fields is undoubtedly a function of their sizes and absolute intensities. Newson (1958) reported that a large inducing field in the extreme periphery of the visual field still exerts a contrast effect. However, the results indicate that the magnitude of the contrast effect decreases rapidly with spatial separation.

Simultaneous Hue Contrast

A colored inducing field induces the complementary color in an achromatic test field. Colored shadows are often cited as a dramatic example of the effects of color contrast. A white screen is illuminated by both yellow lamp light and sky light or white light. An object is placed so as to intercept the yellow lamp light and cast a shadow on the screen. The shadowed area is illuminated by only white light while the remainder of the surface is illuminated by both yellow and white light. The shaded area, illuminated only by white light, will look blue. The yellow light reflected from the remainder of the surface induces a complementary blue color in the shadowed area.

Kinney (1962) has investigated the factors affecting chromatic color contrast. As in the case of achromatic colors, she found that the amount of the complementary color induced in an inlying test field increases as the size of the inducing field is increased, and as the luminance ratio between the inducing field and the test field is increased. She also found a small increase in color contrast with the increased purity of the color in the inducing field. Lightness contrast also occurs with chromatic colors. For example, a yellow test field becomes brown, a red test field becomes maroon, and so on, when the luminance of a surrounding achromatic inducing field is increased. Hue and lightness contrast effects may combine to produce a great variety of contrast colors. For example, Kinney (1962) reported that the colors induced in an achromatic test field by a blue surround varied from a canary yellow to an antique gold to a toast brown as the luminance ratio of the inducing field to the test field was increased. With high luminance ratios, the induced color appears black and desaturated.

Effects of Texture, Contour, and Configurational Factors

Contrast has been shown to vary in a systematic way as a function of the luminance, size, and separation of the inducing field from the test field. Contrast also depends on texture, contour, and figure-ground relationships in ways that cannot at present be systematically interpreted. It appears that the texture of a surface diminishes contrast. For example, contrast effects appear to be greater with completely textureless stimuli (Berman & Leibowitz, 1965). Thouless (1931) also reported that the contrast effect is still perceptible in the afterimage when it is no longer present in direct perception. Presumably, the surface texture and contour are blurred, enhancing the contrast effect. The tissue demonstration also shows how contrast is enhanced when texture and contour are obscured. Viewing a gray paper on a colored background through a thin tissue paper enhances contrast. The phenomena of tissue contrast has been known for a hundred years, but the basis for it is still unknown.

The effect of a contour in minimizing the contrast effect within the contoured area is shown when a gray ring is placed midway on a background half white and half black: the gray ring appears to be of uniform lightness. When a thin line or pencil divides the ring in two along the background division, the half of the gray ring on the white background appears darker than the half of the gray ring on the black background. Berman and Leibowitz (1965) have studied this effect quantitatively and reported that contrast increased as the width of a dividing line between halves of the target was increased. A fine line whose width was only thirty seconds of arc—i.e., the angle subtended by individual cones—was effective in increasing contrast. They interpret their results as indicating the existence of mechanisms that equalize differential contrast effects within the contours of a figure.

O'Brien (1958) found that contrast depends greatly on the luminance transition between areas. Contrast is greater with a sharp contour than with a graded contour produced by a gradual change in luminance. The complex relationship between contours and contrast is strikingly illustrated in the following demonstrations reported by O'Brien (1958). When a 10-degree sector of black was rapidly rotated on a white disk (a in Figure 2), a sharp contour was seen separating the disk's outer region, which appeared lighter than the inner region. When the change in luminance was made gradual by changing the shape of the black sector (b in Figure 2), the change in luminance was not seen and the disk appeared uniformly light without a contour. When a sharp decrement in luminance next to the black sector and a gradual transition away from it to the white sector occurred (c in Figure 2), the inner disk appeared as light or lighter than the outer region of the disk. Thus, modifying the luminance at the contour of a region could make an area of physically lower luminance appear lighter than one of physically higher luminance. An experiment by Krauskopf (1963) further illustrates the importance of contours, showing how they act to determine the color of an area. If an image is stabilized, i.e., keeps stimulating the same retinal receptors, the image will disappear during prolonged viewing. A central disk of one color was surrounded by a ring of another color. The inner contour of the ring was stabilized (i.e., the boundary between the two colors) while the outer contour of the ring was not stabilized. When this occurred, the central disk disappeared and the whole target, disk and ring, took on the color of the surrounding ring.

A demonstration by Wertheimer (Benary, 1924) indicates that figure-ground relationships may modify contrast. If a small gray triangle is placed upon the arm of a black cross and a similar gray triangle is placed within the angle formed by two adjacent arms, the gray triangle spatially included within the contours of the black cross looks lighter than the gray triangle outside these contours (Plate 2). The amount of black and white immediately surrounding the two triangles is the same. However, the included triangle is seen as lying on a black ground and the excluded triangle on a

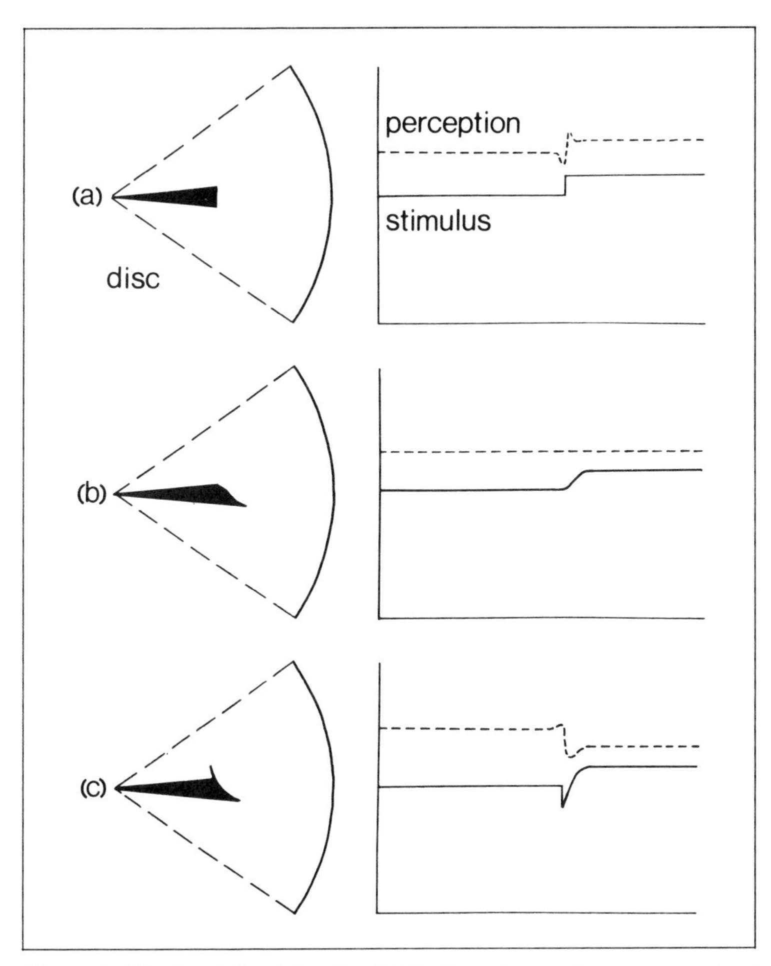

Figure 2. Effects of the intensity distribution at a contour on perceived lightness. (a) Sector; (b) Sector with S-shaped transition; (c) Sector with arrow-shaped transition. (From V. O'Brien, Contour perception, illusion and reality, *J. Opt. Soc. Amer.*, Vol. 48, no. 2 [1958], 118.)

Plate 2. The Wertheimer-Benary demonstration. The gray triangle within the contours of the black cross looks lighter than the gray triangle outside these contours.

white ground. The triangle appears darker when seen against the lighter background. The Wertheimer demonstration has been confirmed with a variety of patterns by Benary (1924) and Mikesell and Bentley (1930). The importance of figure-ground relationships is given additional support by Coren (1969). He reported different contrast effects on a reversible pattern depending on which part of the pattern is seen as figure and which as ground. Greater contrast occurred when a gray area was seen as figure than when the identical gray area was seen as ground. Contrast apparently is also reduced if a surface is seen as standing in front of a background. Recent studies have reported that lightness contrast decreases with increasing separation in depth between a test field and an inducing field (Gogel & Mershon, 1969; Lie, 1969b; Mershon & Gogel, 1970). The findings that figure-ground relationships and apparent depth affect lightness contrast indicate that central factors interact with the processes involved in contrast. It should be noted that Lie (1969b) believes that lightness contrast is independent of the separation in depth of test and inducing fields and that the reported effects are the result of uncontrolled variations in the distribution of light on the retina.

Theoretical Explanations of Contrast

Helmholtz (*1867*, 1925) attributed both lightness contrast and hue contrast to a central transformation in which there is a serial shift in the correspondence of luminance and lightness and that of chromaticity and hue (see p. 48). However, there is strong evidence that supports the theory of Hering (*1874*, 1964) that contrast is produced in large part by processes of neural inhibition and excitation at relatively peripheral levels of processing. Recent electrophysiological studies indicate that the basic lightness contrast relationships are interpretable in terms of lateral neural inhibition (Ratliff, 1962). Von Békésy (1968b), on the basis of psychological experiments, has distinguished between Mach-type and Hering-type lateral inhibition. Mach-type inhibition is associated mainly with the perception of contours and is the basis of contour enhancement effects such as occur in Mach bands. Hering-type inhibition covers a larger area and is the basis of the decrease in perceived lightness such as occurs in simultaneous contrast. De Valois and Pease (1971) have presented evidence that the lateral inhibitory mechanisms found in the retina and geniculate account for contour enhancement, but not for simultaneous lightness or hue contrast.

At present the exact nature of the interactions involved in simultaneous lightness and hue contrast are unknown. We may very briefly indicate three types of explanation that have been proposed. A common theoretical formulation assumes an initial nonlinear transformation that describes the relationship between the light intensity and the resulting neural activity, followed by lateral inhibitory interactions that can be represented linearly (Békésy, 1968a, b; Cornsweet, 1970; Treisman, 1970). If the initial transformation is taken to be logarithmic, the lightness of a test field will remain constant as long as the test field to inducing field ratio is constant; such a model would not account for the Stevens and Stevens (1960), and Jameson and Hurvich (1961) data (see p. 36). A linear model of the lateral inhibition interactions may, however, be applied to a nonlogarithmic transformation; this would account for changes in lightness of a test field when the luminances of both a test field and an inducing field are changed proportionately by the same factor (Cornsweet, 1970). A second type of model does not assume that the luminance of the inducing field directly inhibits the neural intensity signal of a test field. Rather, these models propose that there is a direct response to the luminance difference between a stimulus and its background (Semmelroth, 1970) or to the luminance difference between a stimulus and the average field luminance (Judd, 1941; Helson, 1943; Marimont, 1962). Marimont has presented a model that accounts for the lightness enchancement and depression reported with increasing illumination (see p. 80). The model of Judd represents an up-to-date presentation of Helmholtz's theoretical position (see p. 53). A third type of model is based on the idea of opponent lateral interactions (Hering, *1874*, 1964). The fundamental principle is that a test field shows decreased sensitivity to the color of an inducing field and an increased sensitivity to the complementary color of the inducing field. For the black-white system, the opponent induction caused by an inducing field of higher luminance has the effect of increasing the blackness response in a test field. Jameson and Hurvich (1964) have presented a quantitative description of the contrast interactions occurring in a number of experiments based on an opponent process model (see also Flock, 1970; Jameson & Hurvich, 1970).

There is general agreement that similar physiological mechanisms are the basis of hue contrast. Experiments on the color induced into a test field by inducing fields of different hues (Kinney, 1962) certainly seem to involve lateral inhibition or opponent lateral interactions. In such experiments the color of the inducing field tends to evoke the complementary color in the test field. The demonstration of a colored shadow described on p. 39, taken from Koffka (1935, p. 259), does not fit this rule. Koffka reported that the nonshadowed area, though reflecting yellow and white light, looked white. Helmholtz (*1867*, 1925) also noted that, in the phenomenon of colored shadows, "a very weak colouration of the dominant light elicits quite as distinct contrast colourations as the most saturated" (vol. 2, p. 276). It is possible that the phenomenon of colored shadows differs from simultaneous contrast effects due to lateral neural interactions and involves other factors. The interpretation given to colored shadows is of theoretical importance.

One interpretation is that the phenomenon of colored

shadows involves a central transformation process, as Helmholtz suggested. Helmholtz's explanation of contrast is closely tied to his explanation of constancy; both contrast and constancy involve a change in the stimulus identified as white. According to Helmholtz, white is based on the degree of excitation of the three fundamental sensations (color receptors). He wrote: "The comparison of the intensity of different sensations of colour is extremely uncertain and inaccurate. And, therefore, any determination of white based on such a comparison must be inaccurate too; and pretty considerable variations will be possible in our estimates of white on different occasions, as is actually found to be the case" (vol. 2, p. 274). Within limits, the average color reflected from a scene is identified, according to Helmholtz, as white. In the colored shadow demonstration described previously the white in shadow looks blue because of the false identification of the yellow plus white light as white. A real white, under those conditions, will look blue. Koffka (1935) argued that the colored shadow demonstration is not the result of ordinary simultaneous contrast; for the colored shadow to be the result of ordinary simultaneous contrast, the nonshaded area would have to appear yellow. Koffka's explanation is similar to that of Helmholtz, and explains the phenomenon in terms of the tendency to see the background color in a scene as achromatic or as neutral as possible. It should be pointed out that Helmholtz in support of his explanation cited experiments in which immediate memory modifies the contrast effect in colored shadows (vol. 2, p. 272).

A second interpretation, the commonly accepted view, considers the colored shadow phenomenon an example of ordinary hue contrast. Walls (1960) argued that the effects of simultaneous contrast are independent of adaptation effects. Although the processes of chromatic adaptation may make the yellow light neutral in appearance, the processes of lateral interaction that produce contrast will still induce the blue appearance. The contrast color is complementary to the original stimulus and not to the color that is seen, which is influenced by adaptation. Walls (1960) suggested that Land's (1951) failure to recognize this fact is what has led him to believe that his phenomenon is not explicable in terms of traditional color mechanisms.

Assimilation

Von Bezold (Evans, 1948) illustrated an effect which is the opposite of contrast and has been called assimilation. Colored areas appear lighter when overlaid with white lines than when overlaid with black lines. Von Bezold's figures are illustrated in Evans (1948, p. 192). Assimilation can also occur with achromatic colors: light areas lighten a gray background and dark areas darken a gray background. Newhall (1942) presented an extensive discussion of the phenomenon and reported that assimilation is more variable than contrast. Assimilation appears to be favored by narrow lines as found in outline figures and by casual observation (Newhall, 1942; Burnham, 1953). Helson (1964), conducting the most systematic studies of the parameters affecting assimilation, found that narrow lines and spaces yielded assimilation (the darkening of the gray background by black lines or the lightening of it by white lines), and wide lines and spaces yielded contrast (the darkening of the gray by white lines or the lightening of it by black lines). Assimilation also tends to be greater when the lines and background do not differ greatly in luminance. Helson (1964) has hypothesized that small differences in stimulation in neighboring areas result in summation, yielding assimilation, and that large differences result in inhibition, yielding contrast. Beck (1966) found that

Helson's hypothesis failed to fully account for his data. He reported a marked asymetry when lines of differing reflectance were placed on a gray background. Contrast occurred when the reflectance of the lines was greater than that of the background, and assimilation occurred when the reflectance of the lines was less than that of the background. He also reported that contrast was not affected by repeated judgments, whereas assimilation was reduced. Assimilation may reflect the operation of several different factors. The "graying" that occurs with patterns of small black and white dots, which is the basis for the manufacture of shading materials, is stable and not readily influenced by the manner of observation. Assimilation in this case may result from a physical scattering of light in the eye, and lateral physiological interactions. Another type of assimilation effect that is weaker and less stable may be a function of how an observer views a pattern. It appears to occur when the observer looks at a pattern as a whole. Assimilation is reduced and the observer begins to report contrast when he attends to a given area within a pattern as an isolated feature. A theoretical explanation of assimilation in terms of the effects of attention on lightness perception has been proposed by Festinger, Coren, and Rivers (1970).

Adaptation

Light and dark adaptation refers to the changes in sensitivity which allow the eyes to adjust to changes in ambient illumination. The sensitivity of the eye increases in the dark and decreases with exposure to light. Color sensitivity also changes. For example, exposure to red light decreases the sensitivity of the eye to red light. Chromatic adaptation is a nonlinear function of the physical stimulus and differs for each of the three assumed primaries of the Young-Helmholtz theory (MacAdam, 1961, 1962). Dowling (1967) has called special attention to the fact that there are two types of changes in adaptation. There is a rapid adjustment of sensitivity that is neural in origin, and a slower adjustment of sensitivity that is photochemical in origin. Schouten and Ornstein (1939) found that rapid adaptation effects occur in retinal regions adjacent to a region in which the eye is exposed to a light stimulus. The change in sensitivity decreases with the distance from the light stimulus. Changes in the adaptation are local. Rushton and Westheimer (1962) found that the adaptation produced by a 5-degree target presented to central vision depended upon the total amount of light within local regions of about 1 degree and not upon the distribution of light within these regions. They concluded that adaptation occurs not within single retinal receptors but at or beyond the first retinal level at which spatial summation takes place. Dowling (1967) has speculated that adaptation occurs because of changes in the sensitivity of bipolar cells and that adaptation takes place over the retinal field of a ganglion cell. No experimental evidence appears to exist for the adaptation of the eye as a whole; this adaptation results from the successive local readjustments in sensitivity as the eyes move from one part of a scene to another and is determined roughly by the average color and intensity of the light from a scene.

The effect of adaptation on lightness perception expressed according to a formula derived from Adams and Cobb summarizing how the Weber fraction varies as a function of light adaptation is shown in Figure 3 (Judd and Wyszecki, 1963). The figure shows how perceived lightness varies as a function of the luminance of a target (L), and the luminance of a background (Lb) to which the observer's eyes are adapted. The background is assumed to be large compared to the size

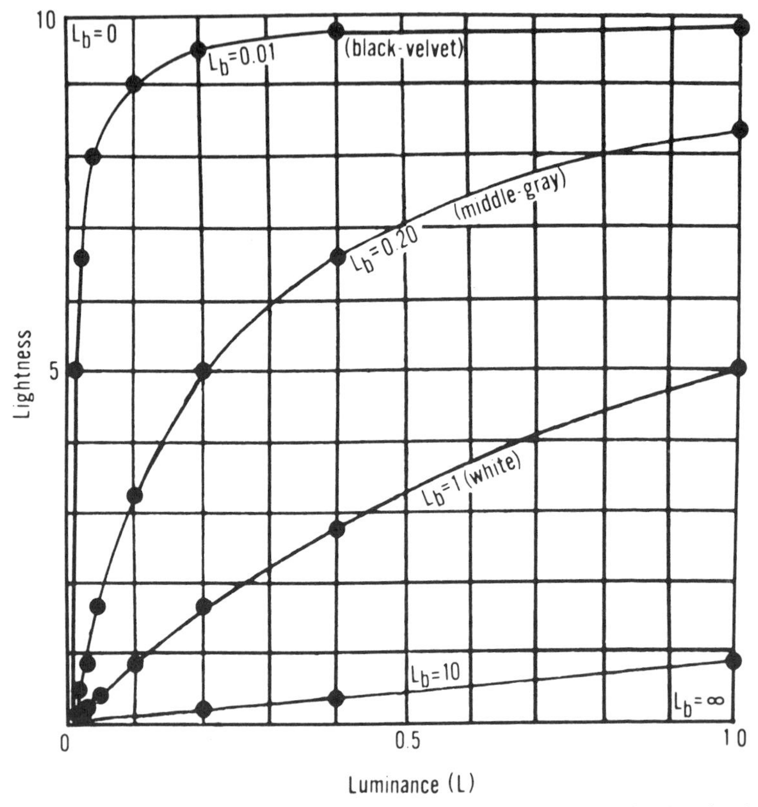

Figure 3. The lightness of a target as a function of the luminance *L* of the target and the luminance *Lb* of the background to which the observer's eyes are adapted. (Adapted by permission of John Wiley & Sons, Inc., from D. B. Judd and G. Wyszecki, *Color in business science and industry* [New York: Wiley, 1963], p. 269.)

of the target. Lightness is expressed on an eleven point scale from zero (ideal black) to ten (ideal white). The luminance of the target ranges from zero (L = 0) to that of a white sample (L = 1) in a normally lighted room. Figure 3 shows that the adaptive state of the eyes modifies the spacing of the lightness scale. For a dark-adapted observer (Lb = .01, a black-velvet background), the gray scale is greatly contracted. At low illuminations, the lower values on a gray scale lose distinctiveness and perceived lightness ranges from white to middle gray. One sees both a black and dark gray as about the same middle gray. For a lightadapted observer (Lb = 1, a white background), the gray scale expands and the targets of low reflectance will be seen as dark gray and black. This scale expansion is also true of the lightness variations with chromatic colors. It is impossible to see brown, maroon, and olive-green distinctly unless they are in good light. If the adapting luminance is ten times that of a white sample on a gray scale (Lb = 10), the lightness scale is again compressed and all samples on the scale appear dark.

Two aspects of Judd's formula should be noted. First, the formula implies that lightness is independent of the illumination. Lightness remains constant if the relation of background luminance to target luminance remains constant. Second, the background luminance not only controls the adaptive state of the eyes but also introduces simultaneous contrast effects. The combined effects of adaptation and contrast greatly complicate the interpretation of the lightness changes illustrated in Figure 3. There are various ways in which the lightness changes can come about. The assumption made in deriving the formula is that lightness perception can be described in terms of the alteration of overall sensitivity of the visual system without specific consideration of how different mechanisms interact. A discussion of the effects attributable to adaptation and to simultaneous contrast may be found in Stevens and Stevens (1960).

Adaptation effects occur in what has been called successive contrast. Successive contrast refers to the continuing influence that a color just previously seen has on the observer's eye; it depends on eye movements that superimpose different stimuli on the same retinal region. Negative afterimages are examples of successive contrast. A negative afterimage is opposite in brightness from the original light stimulus and complementary in hue to it. If one fatigues the red receptors by looking at a red light for some time and then looks at an achromatic light, the achromatic light will be tinged with the complementary color, blue-green. The development of a negative afterimage takes several minutes and is the result of the slower processes of adaptation. Both light-dark adaptation and chromatic adaptation act to preserve the daylight appearance of colors, or color constancy, through neutralizing stimulus changes. For example, an observer will continue to see a white cardboard as white, or almost white, although the illumination on it is changed from daylight to predominantly yellow light mixed with some white light. Chromatic adaptation decreases the sensitivity of his eyes to yellow light and increases their sensitivity to blue light; the increased sensitivity of the blue receptors (i.e., the blue negative afterimage) neutralizes the yellow light coming from the white screen and produces the appearance of a white screen. However, chromatic adaptation is generally not complete and color changes due to changes in the spectral composition of the illuminant are not completely neutralized (Helson & Judd, 1932). Adaptation results in full neutrality only for unsaturated colors or with complete homogeneity of retinal stimulation, as in a *Ganzfeld* (Hochberg, Treibel, & Seaman, 1951; Weintraub, 1964).

The term "adaptation" has been used to refer to two different processes, which need to be carefully distinguished. One is the process that changes the sensitivity of the eye and that probably occurs at relatively peripheral stages (Dowling, 1967). Successive contrast effects are attributable to such sensitivity changes, which we shall call "sensory adaptation." The other process involves a central adjustment of the chromatic neutral point (i.e., the color stimulus which will be seen as achromatic) based on the average color reflected from a scene (Koffka, 1935; Judd, 1940). This second sense derives from psychophysical experiments for which, at present, there is no independent physiological evidence. Color changes occurring in adaptation have been variously ascribed to both processes. Theoretically, it is important to distinguish the effects that are attributable to each (see p. 85).

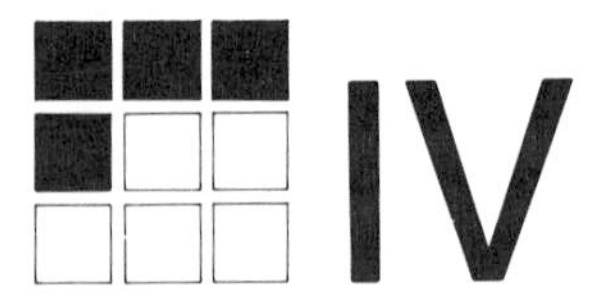

IV

Constancy Experiments, I

The mechanisms of simultaneous contrast and sensory adaptation operate to keep colors constant when the light reflected to the eyes changes. Constancy is favored by the fact that through the operation of simultaneous contrast, the relative luminances stimulating the observer's eyes strongly influence the lightness he perceives. Similarly, the negative afterimage resulting from sensory adaptation tends to neutralize the color change and to maintain the original daylight color. An important question is: Can color constancy be subsumed under the general principles of simultaneous contrast and sensory adaptation? The fact that color often appears to be more constant in everyday conditions than in laboratory experiments suggests that the factors affecting constancy depend on the perceptual setting. An experimental paradigm may exclude certain factors and emphasize others. Therefore, any account of the relation between constancy, contrast, and adaptation should include a consideration of how a particular experimental paradigm may affect the processes that govern perception of surface color. In this chapter we will summarize the results of constancy experiments in a dark room, an air-shadow setting, and separately illuminated fields.

Before beginning the survey of experimental studies, we must define a measure of constancy. Some form of ratio has commonly been used for this purpose. An observer who matches surface colors in accordance with the intensity and quality of the light entering the eyes would be exhibiting zero constancy. An observer who matches surface colors in accordance with their physical reflectance would be exhibiting 100 per cent, or perfect, constancy. The ratio measure expresses an observer's match as a percentage value according to its position on this continuum. There are several ratio measures that have been suggested (Katz, *1911*, 1935; Brunswik, 1928; Thouless, 1931; Landauer, 1962). None, however, is without drawback (Sheehan, 1938; Leibowitz, 1956; Hurvich & Jameson, 1966). The most widely used ratios are those of Brunswik and Thouless.

The Brunswik ratio (B.R.) can be explained most easily by an example. In one chamber a gray disk that reflects 20 per cent of the incident light is illuminated by 10 foot-candles (ft.-c.). We shall refer to this disk as the standard stimulus. In an adjoining compartment a wheel on which papers of differing reflectance have been mounted is illuminated by 50 ft.-c. The observer is able, by turning the wheel, to view one paper at a time. He is given the task of choosing the comparison paper that is equal in grayness to the standard disk. Zero constancy would be exhibited if an observer matched the standard disk to a comparison paper reflecting 4 per cent. Such a match would equate the luminances reflected from the standard and the comparison targets; i.e., both the disk and paper reflect 2 foot-Lamberts (ft.-L.). Perfect constancy would be exhibited if an observer matched the standard disk to a comparison paper of the same reflectance, i.e., 20 per cent. The degree of constancy for different matches is expressed by the formula:

$$B.R. = \frac{J-S}{R-S}$$

where J is the reflectance of the paper judged equal to the standard; R is the reflectance of the standard, and S is the reflectance of the comparison paper whose luminance under the given illumination is the same as that of the standard stimulus. The Brunswik formula expresses the ratio of the difference between an observer's judgment and the match equating the intensities of the light reflected to the eyes, i.e., $J-S$, to the difference between a match equating physical reflectance and a match equating light intensities, i.e., $R-S$. In the above example, R equals 20 per cent and S equals 4 per cent. As indicated, zero constancy occurs when J equals S (4 per cent) and perfect constancy occurs when J equals R (20 per cent). If an observer were to match the grayness of the disk with a paper reflecting 16 per cent, i.e., J equals 16 per cent, the Brunswik ratio would be .75. A ratio of .75 means that an observer's match falls three quarters of the way between a match equating luminance and a match equating reflectance.

$$B.R. = \frac{16-4}{20-4} = .75$$

The Thouless ratio (T.R.) substitutes for J, R, and S their logarithms.

$$T.R. = \frac{Log\ J - Log\ S}{Log\ R - Log\ S}$$

The substitution of logarithms is based on the assumption that lightness varies linearly as a function of the logarithm of gray. The two percepts were, however, not identical: the more intense ring and disk appeared brighter. Wallach (1963) proposed that light stimuli give rise to two different perceptual effects. One effect is dependent on the absolute luminances of the light stimuli and the adaptation of the eyes and determines a bright-to-dim brightness impression. A second effect is dependent on the ratios of light stimuli and determines a white-to-black lightness impression. Wallach (1948) did not specify the nature of the "ratio response" mechanism but has since suggested (1963) that it is the result of interactions between neighboring retinal regions.

A perceptual setting of the type described has also been used by Leibowitz, Myers, and Chinetti (1955), who had observers match the lightness of a 1 degree square comparison target in a dark field to a 1 degree square standard target (reflectance 19.77 per cent) seen against either a white (reflectance 62 per cent), gray (reflectance 19.77 per cent), or black (reflectance .85 per cent) background subtending 14 by 19 degrees. The illumination of the standard target ranged from .00073 to 1,580 ft.-c., a range of over 1,000,000 to 1. With the black and gray backgrounds, Leibowitz et al. concluded that the matches did not differ greatly from those expected based on equating the luminances of the two targets, i.e., zero constancy. With a white background, the lightness constancy of the target was greatly increased. Constancy, however, was not perfect, and as the illumination increased there was also an increase in lightness. The results of Leibowitz et al. were plotted by Stevens and Stevens (1960) and are shown as the filled circles in Figure 1. It can be seen that the increase in lightness with increase in illumination agrees with the general finding obtained by Stevens and Stevens (1960). Dunn and Leibowitz (1961) investigated the effect of an adjacent inducing field on lightness constancy when the intensity of the illumination is changed. In this experiment a 1-degree gray target reflecting 36.2 per cent was matched as a function of its angular separation from a 6-degree inducing field whose reflectance was 86 per cent. The illumination was varied from .056 ft.-c. to 275 ft.-c. They found that constancy decreased as the separation between the target and inducing field increased. When there was a separation of more than 6 degrees the match was based on equating luminance, and very little constancy was exhibited. This is consistent with what would be expected on the basis of contrast. As indicated in Chapter III, contrast effects decrease markedly as the separation between targets increases beyond a few degrees (Fry & Alpern, 1953; Leibowitz, Mote & Thurlow, 1953). The finding by Leibowitz and Chinetti (1957) that decreasing exposure time to .0002 second actually improved constancy is also consistent with the view that in the dark-room setting contrast is a major factor determining the perception of lightness. Dunn and Leibowitz (1961) report that Chinetti found that shorter exposure duration yielded greater contrast effects.

Air-Shadow Setting

A basic paradigm for investigating the constancy of lightness was introduced by Katz (*1911*, 1935) and has been employed with modifications by Helson (1943), Henneman (1935), and Locke (1935). In the experimental arrangement shown in Figure 4, an observer is seated in an illuminated room and faces two matte targets of neutral hue. One target stands in the overall illumination of the room; the other target stands in a spatial region where illumination has been reduced by means of a partition that prevents light from reaching the space behind the partition. This creates an air shadow, i.e., the impression of a tridimensional space of reduced illumination. The two targets are seen as standing in front of a background. Each is textureless and uniformly illuminated, and possesses no characteristic or familiar color. The observer is asked to adjust the lightness of the target in the illuminated region to match the lightness of the target in the shadowed region.

In Katz's (*1911*, 1935, p. 121) original study, a gray cloth served as a common background for two color wheels. A color wheel which was white over its entire 360 degrees served as a standard and was placed in the shadow region. The comparison color wheel was placed in the illuminated region. The observers were seated about 80 centimeters from the color wheels. The experimenter adjusted the sectors of white and black on the comparison disk so that when the disk was rotated, the observer saw the lightness of the comparison disk as matching that of the standard. A setting of the comparison disk to 360-degrees white would represent perfect constancy. The matches of two observers, when illuminances of the two spatial regions had a ratio of approximately 60:1, gave Brunswik ratios of .31 and .25. As the depth of the shadow on the shaded side was increased, Katz found that constancy decreased. However, the apparent darkening of the shadowed disk was not proportional to the decrease in illumination. When the ratio of the illuminances of the two regions was approximately 1.8:1, the illuminated disk with a white sector of 246 degrees and a black sector of 114 degrees matched the lightness of the shadowed disk with 360-degrees white. When the ratio of the illuminances of the lighted and shadowed sides was approximately 90:1, the illuminated disk with a sector of 116-degrees white and a sector of 244-degrees black matched the lightness of the shadowed disk with 360-degrees white. Thus, the darkening of the

considerable individual differences. Katz also found that constancy decreased as the exposure time was decreased. For an illumination ratio of 60:1 (Katz, *1911,* 1935, p. 142), the Brunswik ratio was .15 when the exposure time was 3.4 seconds, and .03 when the exposure time was .6 second.

Helson (1943), using the same experimental arrangement as Katz, varied (a) the ratio of the illuminances in the shadowed and illuminated regions, (b) the reflectance of the standard disk in the shadowed region, and (c) the reflectance of the background. Three illumination ratios were used: 18:1, 42:1, and 1,794:1. Three different backgrounds were used in the main series of experiments. The standard and comparison disks had a common background that was either white, gray, or black. The reflectance of the standard disk in the shadowed region varied in six steps from 360-degrees white to zero or 15-degrees white. As did Katz, Helson found that, though constancy of a white disk decreased as the incident illumination in the shadowed region decreased, the decrease in perceived lightness was more gradual than the decrease in illumination. Helson also found that a black disk became darker under a higher illumination (see p. 36). He reports that the reflectance of the background was important in determining the observer's lightness matches. In general, constancy was greatest with a white background, next with a gray background, and least with a black background. However, a black background sometimes produced almost the degree of constancy found with a white background. For example, with an illuminance ratio of 18 to 1, an observer's matches of a disk of 90-degrees white (i.e., a reflectance of 21.9 per cent) gave a Brunswik ratio of .815 with a white background and .651 with a black background. The reflectance of the white background was 80 per cent and of the black background 3 per cent. Thus, constancy occurred even when ths disk reflectance, 21.9 per cent, was considerably greater than the background reflectance, 3 per cent.

In Henneman's (1935) study the background on the shadowed side was a flat white and the background on the illuminated side a flat black. The shaded white background was equated in luminance to the illuminated black background. A gray curtain that was placed beyond the immediate backgrounds constituted a common background for the standard and comparison disks. The standard target was a flat white disk approximately equal in lightness to the background placed in the shadowed left region. The observer adjusted the lightness of a comparison disk standing in the room illumination to equal the lightness of the white disk in the shadowed region. The observer sat three meters from the test disks. The time of observation was limited to three seconds. Henneman was concerned with the effect of the complexity of a visual field upon lightness constancy. The experiments indicated that lightness constancy was greatest when the gray background beyond the immediate background was visible and when a number of smaller darker disks were present. The Thouless ratio (a) with monocular vision under complete reduction (vision was restricted to the areas of the standard and comparison disks) was .03; (b) with binocular vision with the homogeneous fields around the shadowed and comparison disks visible, .29; (c) same as (b) but with the remote (room walls) background visible, .44; (d) with the same observation conditions as (b) except that three small disks were placed in the region of the standard disk, .50; and (e) with the same observation conditions as (d) except that the remote background was visible, .53. When the distance of observation was decreased to .75 meters and the field conditions were the same as in (c), the Thouless ratio was .53. Thus, as in the results of Helson and Katz, the presence of reflectances in the field that were lower than the reflectance of the standard improved constancy. As did Katz, Henneman found that decreasing the exposure time from 3 seconds to .75 second reduced constancy by a statistically reliable amount.

Henneman (1935) also investigated whether instructions to an observer could change his lightness matches. Observers were divided into two groups according to whether in preliminary tests they had achieved low or high Thouless ratios, i.e., had failed to exhibit or had exhibited constancy of lightness perception. The average Thouless ratio for the low group was .22 and for the high group .72. Henneman instructed the observers in the low group to adopt an object-directed attitude in order to see the true lightness of the disk. Observers in the high group were instructed to adopt a subjective attitude directed to the intensity of the light coming to the eyes. After instruction, the Thouless ratio of the originally low group became .58 and of the originally high group .47. Observers' reports indicated that the perceptual situation was ambiguous; some said that they could make either of two color matches, "taking the standard as 'gray' and ignoring the illumination of the field, or seeing it as 'white' in a region of shadow" (Henneman, 1935, p. 65). Katz (*1911*, 1935), MacLeod (1932), Helson and Jeffers (1940), and Evans (1948) have also reported that the different attitudes that observers may adopt influence their color judgments. Landauer and Rodger (1964) found that lightness matches varied depending upon whether observers were instructed to equate two disks in different illuminations in terms of reflectance (same gray appearance assuming the targets are illuminated equally), brightness (same absolute luminance), or lightness (same gray appearance). Landauer and Rodger reported that observers receiving instructions to equate the disks in terms of appearance responded in a characteristic way that clearly differed from the way that observers receiving reflectance and luminance instructions responded.

Locke (1935) compared the lightness constancy of five Rhesus monkeys and five human adults under similar stimulus conditions. The monkeys were first taught to respond to the white target in a black-white discrimination task. The room illumination was then arranged to create a shaded white target and to illuminate the black target directly. The illumination on the black target was then increased and the values at which the animals switched their response from the white to the black target were used to calculate a measure of constancy. The complexity of the visual field was also varied. Complexity was increased by adding black disks in front of the black target and gray disks in front of the white target. Constancy for both monkeys and humans improved with field complexity. Average Brunswik ratios for the monkeys with (a) white and black targets alone was .24, (b) the addition of three darker disks placed in front of the white and black targets but with the field free of shadows, .34, and (c) the addition of three darker disks which cast shadows on the white and black targets, .56. The corresponding ratios for the humans were .07, .14, and .15. Thus, both monkeys and humans showed the greatest constancy for the most complicated field and the least constancy when presented with the white and black targets alone. The difference between the responses to the most complicated field and one in which the targets were presented alone was statistically reliable for both humans and monkeys. In all conditions the monkeys exhibited greater constancy than did the humans. Locke suggests that the difference between the monkeys and human observers is that a monkey's set is to perceive object characteristics rather than isolated color "qualities."

Separately Illuminated Fields

Hsia (1943) investigated constancy when the standard and comparison fields were separately illuminated. In one compartment a standard gray was placed under fixed illumination, while in a second compartment a comparison target was placed under a different illumination. The observer's task was to equate the lightness of the comparison target to that of the standard. The illumination of both targets was varied by raising and lowering overhead lamps. Hsia took great care to eliminate contrast effects. Both compartments had black backgrounds and neither of the compartments was directly illuminated by the overhead ceiling lamps. The side walls of the compartments were covered with matte gray paper that reflected 17 per cent of the incident light. Hsia tested for possible contrast effects from the walls and floor within the compartments but could not detect any. In one experiment, the standard was a gray of 12 per cent reflectance that was illuminated by .75 ft.-c. In this experiment the illumination of the comparison field was always greater than that of the standard field and varied from 3.40 ft.-c. to 29.80 ft.-c. Observers, it was found, exhibited a fair degree of constancy. For example, the reflectance of the comparison target equated to the standard was 9 per cent when the illumination of the comparison field was 3.4 ft.-c. and 5 per cent when the illumination of the comparison field was 29.80 ft.-c. The Brunswik ratios for these matches were .29 and .64. The greater Brunswik ratio at the higher illumination of the comparison field indicates that the mean of observers' matches at the higher illumination was closer to the value that would equate target reflectances than the mean of observers' matches at the lower illumination. The Brunswik ratio at the higher illumination fell at a point 64 per cent of

and a black. The observers were instructed to scan the targets without fixating too long on any single target. They made absolute judgments of the hue, lightness, and saturation of each target. Judgments of hue were made on a sixteen-step scale, and judgments of lightness and saturation on a ten-step scale. Observers made their judgments after their eyes had adapted to the general illumination. The perceived hue, saturation, and lightness of the papers were found to be a function of the relation of their reflectances to the general adaptation level. The adaptation level was found to be a weighted function of all the reflectances in the visual field. Helson (1938) formulated the following principle: Samples whose reflectances are above the adaptation level increasingly take on the hue of the illuminant; samples below the adaptation level increasingly take on the hue of the complementary afterimage of the illuminant, and samples at the adaptation level appear achromatic. This finding typically held for chromatic as well as achromatic samples. Thus, both the reflectance of a colored paper and its perceived lightness under daylight are more important than its perceived hue under daylight in determining its appearance under highly chromatic illumination. Helson found that the background, by controlling the adaptation level, could determine whether a stimulus is seen in the illuminant hue, in the hue of the complementary afterimage, or as achromatic. For example, a medium-gray paper with a white surround, illuminated by red light, will take on the complementary color of blue-green because its reflectance is below the adaptation level. The same paper with a gray surround will appear either achromatic or greatly reduced in saturation because its reflectance is now close to the adaptation level, and with a black surround it will appear saturated red because its reflectance is now above the adaptation level. Helson reported, however, that if a small amount of lamp light is added to the highly chromatic illumination, constancy greatly improves and stimuli will retain their daylight colors. For example, achromatic gray stimuli retain normal color (Helson, 1938) when the intensity of the lamp light (2850°k) is only 7 per cent of the mixture. The results are similar for colored stimuli. Colored stimuli tend to keep their daylight hue if their dominant wavelength is present even in a minor amount in the illuminant. Helson and Jeffers (1940) indicated that an observer's attitude can affect his color judgments. Observers who were set to observe as many daylight hues as possible showed greater constancy with chromatic illuminants than did observers who were not. Judd (1940) reported an experiment similar to that of Helson and Jeffers (1940). Six observers made judgments of the hue, saturation, and lightness of the same Munsell samples used by Helson and Jeffers under daylight and under four chromatic illuminants. The results are similar to those of Helson and Jeffers (1940).

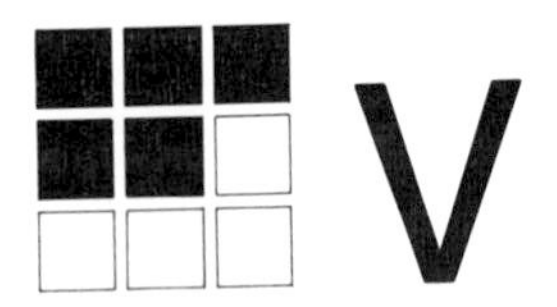

V

Explanations of Constancy

Constancy occurs because there are processes that enable an observer to perceive a color as the same though the stimulus has changed. In this chapter, four approaches that have been taken to the problem of color constancy are described. The results of the experiments reviewed in the preceding section will be considered in relation to these approaches.

Contrast Explanation

The contrast explanation of constancy does not deny that one ordinarily has an impression of the illumination or that colors may have diverse modes of appearance. It asserts, however, that the perception of lightness is determined by sensory processes. Thus, no special consideration need be given to an observer's taking into account either differences in illumination or in mode of appearance—for example, to such differences as those between daylight and chromatic illumination, and between film and surface colors. The relationships between the perceptions of color, illumination, and mode of appearance are a function of their stimulus correlates. According to the contrast explanation of constancy, the covariation of lightness and illumination in Gelb's

(1929) experiment and that of lightness and mode of appearance in the Hering shadow demonstration are parallel phenomena resulting from the covariation of their stimulus correlates (see p. 117 and p. 127).

The results of constancy experiments in a dark-room setting are in accord with what would be expected on the basis of contrast: (a) approximate constancy occurs when the background reflectance is greater than the reflectance of the target; (b) constancy is absent when the reflectance of the background is less than that of the target; and (c) constancy decreases with increased spatial separation between a target and a field of higher illumination. The finding by Leibowitz et al. (1953) that a target surrounded by a large background of reflectance equal to that of the target exhibits some constancy is consistent with the results of the lightness contrast experiment by Torii and Uemura (1965) in which increasing the size of an inducing field of luminance equal to that of a test field depressed the lightness of the test field. The similarity in the results of these experiments is to be expected since the basic experimental conditions of constancy experiments in a dark-room setting (Chapter IV) and of contrast experiments (Chapter III) are highly similar. In the contrast experiment the ratios of the luminances of a test field and inducing field are varied. In the constancy experiment the incident illumination is varied. Variation of the illumination alters the absolute luminances of the target and its surround but keeps constant the luminance ratio between the target and surround. The experiments by Stevens and Stevens (1960), Jameson and Hurvich (1961), and Wallach (1948) can be viewed as either experiments on lightness constancy or on contrast. How such experiments are classified depends on whether one focuses on the change in lightness with changing illumination or whether one focuses on the lightness of a target as a function of the different luminances of the surrounding field.

Simultaneous contrast is important: constancy is typically greater in the air-shadow paradigm for targets whose luminance is below that of the background than for those whose luminance is above that of the background. The reduction of constancy to simultaneous contrast, however, is not consistent with three differences between the results of constancy experiments in a dark-room setting in which the standard and comparison surfaces are perceived to be illuminated uniformly and the results of constancy experiments in an air-shadow setting in which the standard and comparison surfaces are perceived to be in differently illuminated spatial regions.

First, simultaneous contrast is primarily an inhibition of a field of lower luminance by a neighboring field of higher luminance. Thus, constancy in a dark-room setting has been found to be minimal when the luminance of a target is greater than that of the background. However, significant constancy has been found under these conditions in the air-shadow setting. The results in Helson's (1943) experiment are closer to matches based on constancy than to luminance matches, even when the luminance of a target was considerably above the luminance of the background. Further, both Henneman (1935) and Locke (1935) found that constancy increased when a number of smaller, darker disks were placed in the field near the target. Henneman also found that the presence of a remote background beyond the immediate background improved constancy. It should also be noted that Hsia (1943) and Oyama (1968) found constancy under conditions in which the experimental arrangement would minimize contrast effects.

Second, the influence of attitude, or set, is more evident in

viewed in an illuminated space is, it seems, influenced by his adjustment to the prevailing illumination. The increased constancy in the air-shadow setting with an immediate surround of very low luminance compared to the dark-room setting can perhaps be explained by the additional surfaces that have been included in the air-shadow setting; these surfaces establish an impression of illumination that influences the lightness judgments. The individual differences reported under these conditions of observation would reflect basic differences among individuals in their observational attitudes. Some observers may naturally adopt an attitude allowing for the illumination. Instructions may also be expected to influence an observer's set for "taking into account the illumination". Evans (1948) reports observations made with the experimental arrangement used by Hess and Pretori to investigate contrast. In the original Hess and Pretori experiment, observers "viewed" the standard and comparison targets and their surrounds in a room which was completely dark except for the contrast-inducing display. Under these conditions of observation Evans replicated Hess and Pretori's findings (see p. 35). When the standard and comparison targets were seen in a not wholly dark room, the perceived lightness of the comparison target depended on whether the standard and comparison targets were perceived to be uniformly illuminated or not. When observers were instructed to see the standard and comparison targets as uniformly illuminated, the comparison target was seen as darker than when the observers were instructed to see the standard target as illuminated and the comparison target as shaded.

Although the experiments reported in Chapter IV suggest that constancy is not reducible to the sensory processes of contrast and adaptation, no definite conclusions can be drawn because the experiments cannot be interpreted unambiguously. For example, the increase in constancy that occurs when the distance of observation is decreased may be a result of the minute irregularities that become visible on even a smooth surface and provide a basis for the separation of illumination from surface lightness. Texture is an important attribute of the surface mode of appearance that the film mode lacks and appears to be a major factor in separating the impression of illumination from that of the lightness of a surface. In general, the luminances of the background surfaces in the experiments described were not uniform and so could provide the observer with cues to the illumination. Alternatively, one may argue that the presence of even slight luminance gradients may introduce contrast effects through "triggering mechanisms" (Helson, 1943). Moreover, what is ascribed to a difference in attitude of observation may actually be the result of differences in successive contrast due to the way in which the eyes of an observer scan a display. It should be noted, though, that Henneman (1935) and Hsia (1943), considering both their own data and experiments in more complex settings (described in Chapter VII), conclude that constancy is not reducible to simultaneous and successive contrast. More recently, Kozaki (1963, 1965) and Oyama (1968) have interpreted their results as being inconsistent with a contrast explanation of lightness constancy. Further examination of the adequacy of a contrast interpretation of lightness constancy can be found in Freeman (1967).

Adaptation Level Explanation

Rather than attempting to explain constancy in terms of specific sensory interactions, Helson (1938, 1940, 1943) has proposed that colors arise as gradient phenomena with respect to a single reference level which he calls the adaptation level. The adaptation level, operationally, is the stimulus which elicits a perception of medium gray. Quantitatively, the adaptation level has been found to be a weighted function of the luminances and colors reflected from a scene. Helson's adaptation level principle is based on a generalization of the results of his experiments described previously (see p. 70). The principle asserts that in every viewing situation an adaptation level is established; this level is such that stimuli above it take on the hue of the illuminant, stimuli below it take on the hue of the afterimage complementary to the hue of the illuminant, and those at the adaptation level appear achromatic. According to Helson, the adaptation level generalization incorporates, in a single principle, effects which are usually separately ascribed to the sensory processes of contrast and adaptation and to constancy mechanisms. As in the contrast explanation of constancy, Helson argues against theoretical views that take explicit account of either the factor of illumination or mode of appearance in accounting for the perception of surface color.

Whether stimuli will keep their daylight color with changes in illumination depends on many factors. The occurrence of color constancy may have different bases. For example, gray samples will remain achromatic when as much as 93 per cent red light is added to incandescent lamp light (Helson, 1938). The fact that constancy occurs, and light grays are not seen in the color of the illuminant and dark grays in the afterimage complementary hue, as might be expected, would be ascribed to a compensatory adjustment that keeps the adaptation level constant. Presumably, compensation is the result of peripheral adaptive processes that reduce the sensitivity of the red receptors and probably also of central adaptive processes that reduce the effectiveness of the physically stronger red light. When the illumination becomes increasingly monochromatic, the adaptation level can no longer be kept constant through adjustments in sensitivity and shifts toward the illuminant. Constancy, however, may still occur because of the relationship of the stimulus to the adaptation level. Thus, in highly chromatic red illumination a sample of high reflectance which is red in daylight will retain its daylight hue because its reflectance of the illuminant is high; a daylight green of low reflectance will retain its daylight hue because its reflectance of the illuminant is low; and a gray paper will remain achromatic, provided its reflectance is close to the adaptation level. Selective samples may also keep their daylight appearance under highly chromatic illumination (that is not, however, homogeneous) because perceived hue is a function of the hue resulting from the shift in the adaption level and the light reflected from the sample. For example, a dark green of medium saturation when illuminated by red-orange light on a white background takes on the complementary hue of the illuminant and is perceived as a green-blue. The addition of 8 per cent tungsten light, however, causes the stimulus to regain its green appearance. The added yellow light from the tungsten lamp apparently reduces the blue of the afterimage and the observer again sees the daylight green (Helson & Jeffers, 1940). A point that Helson makes should be noted: the concept of adaptation level requires much faster adaptive processes than those that occur in the development of a negative afterimage (Helson, 1943). As indicated, recent experiments have found adaptation effects that may occur instantaneously (see p. 51).

Lightness contrast is a function of the ratio between the luminance of a sample and the adaptation level. A background of low reflectance induces the illuminant color in a target and lightens it; a background of high reflectance induces the complementary color in the target and darkens it. Lightness constancy is perfect if the luminance ratios of a target to the

the effects of discounting the illumination color from the effects of simultaneous and successive contrast. An observer discounts the color of the illumination in the surface mode of perception because the pattern of reflectances in a scene provides him with information about the illumination color. When surfaces are perceived to be uniformly illuminated, Judd proposes that the average chromaticity of the light reflected from a scene is ordinarily identified by the observer with the illumination chromaticity. The discounting of the chromaticity of the illumination by an observer is interpreted in terms of a shifting of the achromatic point in the chromaticity diagram, i.e., the chromaticity, expressed in colorimetric units, which is seen as achromatic. Similarly, the discounting of the intensity of the illumination by an observer is interpreted in terms of a shifting in the lightness-intensity correlation, i.e., the average luminance is seen as a medium gray; higher luminances are seen as increasingly white; and lower luminances are seen as increasingly black.

The axes x and y in Figure 5 are called "chromaticity coordinates" and specify the color quality of the light stimulating the eyes in terms of trichromatic coefficients. Two physically different spectral energy distributions will determine the same color under standard viewing conditions if they have the same chromaticity coordinates. The curved line shows the locus of the spectrum colors in the chromaticity diagram, i.e., the red to violet colors produced by monochromatic wavelengths. The straight line connecting the extremities of the spectrum locus (400 to 700 mμ) represent the colors from blue-purple to red-purple and are produced by mixing in suitable proportions violet light with red light. Colors that lie within the chromaticity diagram are produced by mixing a spectrum color or a purple color with achromatic light. The point C near the center represents the

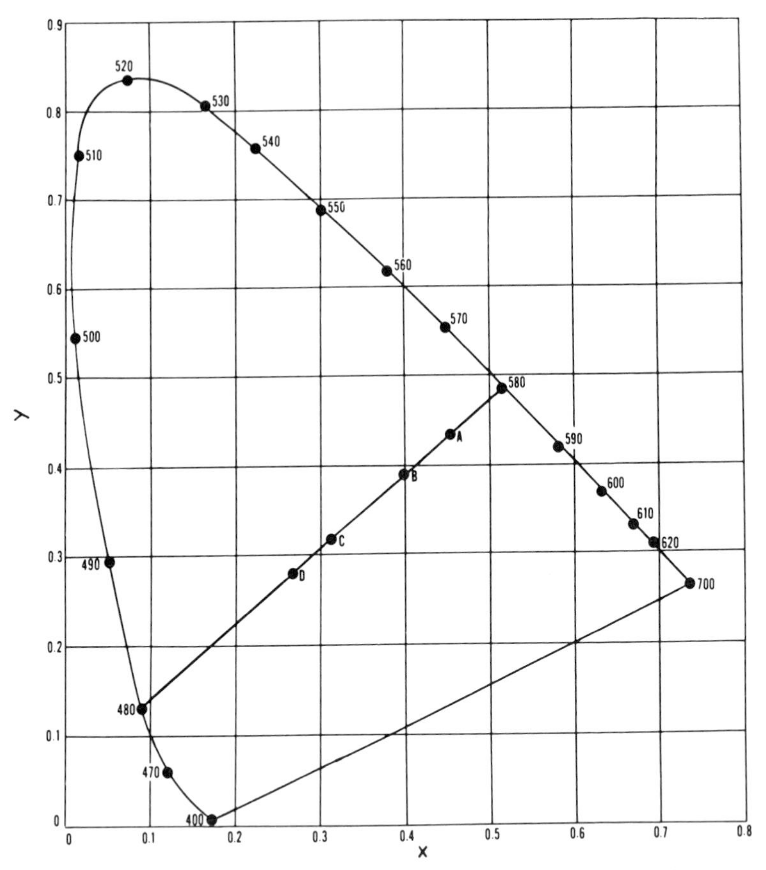

Figure 5. Chromaticity diagram in terms of the trichromatic coefficients.

color quality of the light reflected from an achromatic sample illuminated by daylight. The representation of a color in terms of its dominant wavelength and purity is obtained by extending a line from C through the point representing the color to the spectrum locus. The intersection of the line with the spectrum locus gives the dominant wavelength. The colors plotted at points A and B have a dominant wavelength of 580 mμ(yellow) and the color plotted at point D a dominant wavelength of 480 mμ(blue). Excitation purity is the ratio along the line C of the distance between the achromatic color C and the point representing a color to the distance between the achromatic point C and the dominant wavelength of the color. Excitation purity varies from zero for the achromatic point to one for a spectrum color. Saturation is an increasing function of excitation purity. Under standard viewing conditions the color at A represents a more saturated yellow than the color at B. The color at D represents a desaturated blue. Suppose now the illumination is changed from blueish daylight to yellow-white tungsten light. The achromatic sample that was previously located at C now reflects light that locates it at A and the desaturated blue sample that was located at D now reflects light that locates it at B. If no compensation occurs for the change in the color of the illumination, the samples that previously were seen as achromatic and desaturated blue will now appear yellowish. If the observer discounts the color of the nonneutral illumination, the achromatic point shifts to A, and the original achromatic sample will continue to be seen as achromatic. The point B, which lies on the line between the new achromatic point A and the spectral wavelength of 480 mμ, will continue to be seen as a desaturated blue. If a nonneutral illumination were completely discounted, all colors would retain their daylight hues and color constancy would be perfect.

In practice, color constancy is not perfect. Samples with luminances above the average luminance of a scene take on the hue of the illumination color and samples with luminances below the average luminance of a scene take on the hue of the afterimage complementary to the illumination color. To explain these departures from constancy, Judd takes into account the Helson contrast principle described previously. He ascribes the Helson effect to peripheral successive contrast. The color of a sample is modified by the projection of an afterimage of the average field chromaticity. The formulas of Judd are based on the idea that an afterimage of the average field chromaticity projected onto a sample color of the same chromaticity whose luminance is three-sevenths of the average field luminance will be seen as achromatic. The full development of the negative afterimage takes about five minutes. The Helson effect, therefore, is assumed to take time to develop and to depend on the pattern of an observer's fixations (Judd, 1940, p. 305).

The implications of Judd's theoretical position need to be clearly stated. According to Helmholtz's principle, diffusely reflecting achromatic surfaces perceived to be uniformly illuminated in strongly chromatic illumination appear achromatic though they reflect chromatic light to the eyes. Constancy occurs because an observer discounts the average chromaticity of the scene as the illumination color. Thus, if gray papers are suddenly illuminated by yellow light, the papers will continue to appear gray. The yellow color may be seen, but it is perceived as belonging to the illumination and not to the papers. Koffka's (1935) position is similar to that of Helmholtz. He suggests that all nonselective surfaces will appear achromatic in chromatic illumination because the chromaticity of the background color determines a "color framework" that will appear as neutral as the conditions will allow. Other conditions being equal, the colors of surfaces in a scene are a function of their deviations from the achromatic point established in the chromatic illumination. According to Koffka, the shift in the achromatic point in chromatic illumination from the normal or daylight achromatic point is in accordance with the *Law of Prägnanz*. The new achromatic point is such as to minimize color changes in a scene with the change in illumination.

If an observer looks for several minutes at gray papers under yellow light, the intensity of the light reflected becomes important in determining the color that he perceives. The process of peripheral successive contrast changes his perception. The color seen depends on the amount of yellow light reflected and the strength of the blue afterimage determined by the luminances in the field. A black paper appears as a saturated blue because black paper reflects little yellow light and the blue afterimage predominates. Gray paper reflecting sufficient yellow light to neutralize the afterimage will appear neutral. A white paper will appear a saturated yellow because it reflects enough yellow light both to neutralize the blue afterimage and give the surface a yellow appearance. According to Judd's formulation, neutral papers presented in chromatic illumination would continue to appear achromatic if the effects of successive contrast were minimized. Judd attributes Helson's finding that neutral papers in chromatic illumination take on the hue of the illuminant when their reflectances are above that of the adaptation level and of its complementary afterimage when their reflectances are below that of the adaptation level (see Chapter IV) to the effects of successive contrast, which Helson's experimental conditions did not minimize.[4] Helson's experiment thus

[4] Weintraub (1964) reported that the color of a post-adapting field following adaptation to a red *Ganzfeld* (observers reported seeing brightness but no hue) is a function of the purity but not the luminance of the post-adapting field. Post-adapting fields of equal or greater purity than the adapting field were reported red. Post-adapting fields of purity less than the adapting field were reported as a complementary blue-green. Luminance of the post-adapting field had little effect on hue. According to Judd's hypothesis, it is possible that periph-

represents a failure of constancy because of the effects of peripheral successive contrast. Under these conditions it is not surprising that Helson reports no difference between the perception of color in the film and surface modes.

The equations formulated by Judd (1940) to predict perceived surface color are most easily illustrated graphically. Figure 6 illustrates the color that it is predicted an observer will see if he looks at an array of Munsell colored papers against an achromatic background for five minutes or more in red light. It is assumed that the observer continually scans the field and that he fixates only momentarily a particular color. The dashed line represents a lightness scale projected onto the uniform chromaticity triangle and shows the locus of achromatic points as a function of surface lightness. According to Judd's formulation, a white surface would fall in the neighborhood of the open circle marked W, and a black surface in the neighborhood of the open circle marked BK. The perceived hue of a surface is correlated with the direction of the vector connecting the point representing the lightness of a surface with the point on the uniform chromaticity triangle representing the chromaticity of the light reflected from the surface. The diagram in the upper left part of the figure shows the correlation between hue and the direction of the vector. Saturation is determined by the

---

eral successive contrast effects are strongly dependent on the presence of nonuniform stimulation. Adaptation effects in a *Ganzfeld* in which the entire retina receives uniform stimulation may be the result of a color transformation in which there has been a shift in the achromatic point. Adaptation effects in a *Ganzfeld* would thus be a function only of stimulus purity. It should be pointed out that Helson has suggested that the difference between his results and Weintraub's is that the luminance differences in Weintraub's experiment were not sufficient to yield successive contrast effects (Avant, 1965, p. 251).

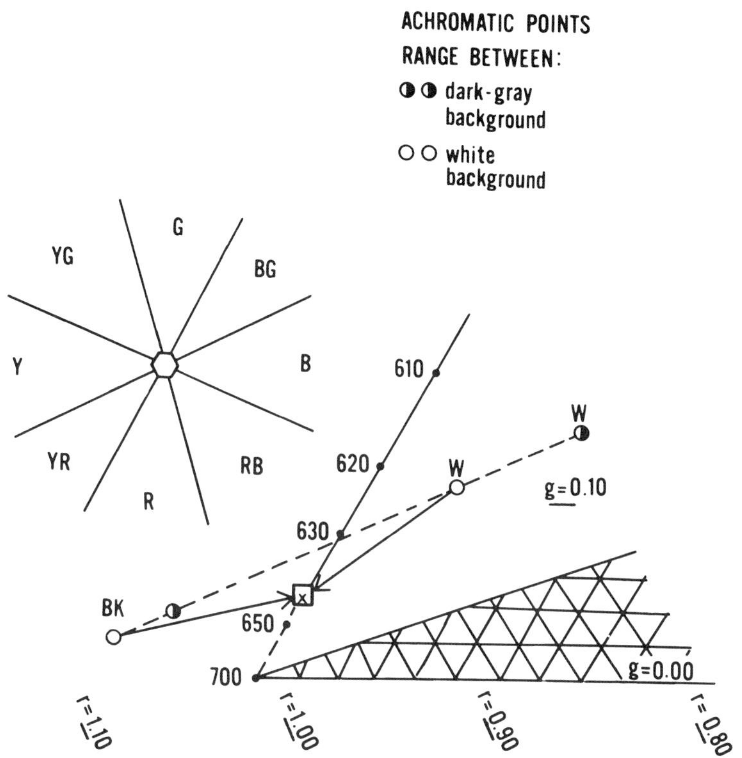

Figure 6. Formulation for the prediction of surface color, for surfaces seen under red light by an observer who has been exposed to the field for five minutes or more. (From D. B. Judd, Hue saturation and lightness of surface colors with chromatic illumination, *J. Opt. Soc. Amer.*, Vol. 30, no. 1 [1940], 16.)

length of the vector. The coordinates of the illumination chromaticity are shown in Figure 6 by a square. The chromaticity of the light reflected by the nonselective papers is the same as that of the illumination chromaticity and is indicated by the x shown in the square. The vector from a white surface indicates that the surface would be seen as a yellow-red. The vector from a black surface indicates that the surface would be seen as a blue. As the lightness of surfaces goes from black to white all possible colors between yellow, red, and blue are encountered. For example, a gray surface whose lightness falls on the dashed line just above the illumination square would look gray with a slight red tinge. The vector connecting it to the x would point downward and be very short.

Since the average field chromaticity and luminance are identified with the illumination chromaticity and luminance, the formulas of Judd may be interpreted in terms of changes of the overall sensitivities of the color encoding mechanisms without reference to "taking into account" the illumination. The sensitivity of each of the independent color mechanisms, which may be assumed to vary inversely with its degree of excitation, determines the adaptive state of that mechanism. Reinterpreted in terms of the state of adaptation of the color encoding mechanisms as influenced by the field chromaticities and luminances, the positions of Judd and Helson coalesce. The Helson contrast effect may be incorporated by assuming that the neural mechanisms that control sensitivity changes alter sensitivity in proportion to the change of illumination when the sample luminance is equal to the average field luminance, alter sensitivity in greater proportion than the change of illumination when the sample luminance is below the average field luminance and alter sensitivity in lesser proportion than the change of illumination when the sample luminance is greater than the average field luminance (Richards & Parks, 1971). The Judd formulas interpreted in terms of the state of adaptation of the visual system are often described as the Helson-Judd theory of color conversion.

The difference between the two interpretations, of course, arises when surfaces are perceived to be nonuniformly illuminated. The Judd formulas are fairly successful in describing the perception of color when surfaces are perceived to be uniformly illuminated (Judd, 1940; Pearson et al., 1969). For uniformly illuminated surfaces, the formula for lightness fails in two principle ways. First, it fails to account for the changes in lightness that occur with changes in the overall illumination (Stevens and Stevens, 1960; Jameson and Hurvich, 1961, see p. 36). As pointed out, the Judd formula implies that lightness remains constant as long as the ratio of the sample luminance to the average field luminance is held constant. Secondly, the formula fails to account for the fact that perceived lightness changes rapidly with changes in sample luminance when the sample luminance is close to the background luminance (Takasaki, 1966). The formulas, however, do not accurately predict color perceptions when surfaces are not perceived to be uniformly illuminated. The application of the Judd formula to Helson's experiment (1943) by Semmelroth shows that the formula fails to predict the perception of lightness when the standard surface is in a shadowed space and the comparison surface is directly illuminated (Semmelroth, 1970). The standard surface in shadow is seen as too light. It should be emphasized that according to Judd the formulas are not applicable when surfaces are perceived to be illuminated nonuniformly. The average luminance in a scene can then no longer be identified with the illumination intensity. When surfaces are nonuniformly illuminated, as when one side of a white cube is more strongly illuminated than another, the formula will not accurately predict perceived lightness, and explicit consideration must be given to illumination cues and object cues. Because of these, an observer will ordinarily see the two sides of a cube as white and unequally illuminated rather than as a cube with white and gray sides.

Judd (1940) found that his formulas predicted 97 per cent of lightness judgments, 84 per cent of hue judgments, and 73 per cent of saturation judgments. These predictions refer only to experiments in which an observer has remained in an enclosure scanning color samples for five minutes or longer. More recently, he has applied his formulas to the Land experiments (Judd, 1960). Land (1959) has recently published a series of experiments describing striking color effects. He has shown that the projection in register of two black and white positive transparencies of the same scene can reproduce the colors in the original scene. The two transparencies may be prepared by taking two photographs of the same scene, using black and white film, one through a red filter (the long record), and one through a green filter (the short record). If red light is projected through the "red" transparency and an incandescent lamp light through the "green" transparency, the colors approximate those in the original scene. On the basis of his experiments, Land concluded that a radical revision of color theory was necessary and has proposed a new theory (Land, 1964). Judd (1960) computed the colors predicted by the formulas he published in 1940 for two of Land's experiments and found the predicted colors to be in good agreement with Land's results. The major contradiction is that Land reported as light blue what Judd had computed as light green. Judd suggests that this discrepancy can perhaps be attributed to the effects of memory color. Belsey (1964) reports that colors in the Land phenomena are displaced toward the memory color of familiar objects. The Judd formulas applied to the Land two-color projection situation were tested by Pearson et al. (1969) and on the whole received good support.

It should be stressed that Judd's explanation takes into account Katz's distinction between the film and surface modes of appearance. Land's (1951) experiment number 20 showed that when two color projection slides are presented out of register the colors seen are few and unsaturated. When the slides are brought into register, the "full gamut of color snaps into appearance." Judd (1960) says:

> When the two separation positives are presented far out of register, the viewer does not perceive any objects, but merely a pattern of light patches; so his report is influenced by successive and simultaneous contrast alone. . . . When the images are brought into registration, the pattern of lights is perceived as representing objects illuminated by some kind of light. The perceived color of this light is discounted and . . . as Land puts it "the full gamut of color snaps into appearance." [p. 258].

A detailed consideration of Land's results and conclusions from a somewhat different point of view from Judd's will be found in Walls (1960), and Wheeler (1963). Both, however, conclude that Land's results can be interpreted in terms of contrast mechanisms and do not require a major revision of color theory.

Alternative Interpretations

The hypothesis that an observer discounts the illumination color in a scene assumes that the discounted value acts as a reference with respect to which his perceptions of the surface colors in the scene are then determined. According to Judd's (1960) formulation, the discounting is automatic and is not easily affected by an observer's attitude or knowledge. Perceptual mechanisms may, however, affect the perception of color in another way. A perceptual stimulus may be ambiguous and allow for alternative interpretations as illustrated by Judd's analysis of another of Land's experiments. Land reported that when *no* transparency is placed in the red light projector, and incandescent lamp light is projected through the green transparency, the observer sees a wash of red light over the black and white image from the incandescent projector. Judd repeated the experiment, and found that the scene could be viewed in two ways: (a) the dark grays could be perceived to differ in chromaticity from the light grays; and (b) both the dark and light grays could be perceived as without hue, through a film of red light. In the second way of seeing the stimulus, constancy occurs and the grays are seen as gray. Constancy, however, does not occur because the observer discounts the color of the illumination in perceiving the surface color, but because he perceives an intervening film of red light.

Since the light reflected to the eyes is jointly determined by the reflectance of the surface, the spatial position of the surface, and the amount and direction of the illumination, there is no fixed relation between the amount of light reflected and a specific surface lightness. Chapter VII reviews experiments showing that the perceptions of surface color, surface illumination, and the apparent spatial position of a surface are interdependent. As for the perception of lightness, there are many experiments which cast doubt on the hypothesis that this interdependence results from an automatic discounting of the intensity of the illumination (see Chapter VI). A different explanation of this interdependence is that the perception of surface lightness involves two components: (a) sensory processes of transduction, enhancement, and abstraction, such as adaptation, contrast, and contour formation, that determine a central neural pattern in accordance with the peripheral luminance distribution; and (b) schemata for the perception of visual surfaces that encode a central neural pattern in terms of lightness differences, differences in surface orientation, or differences in surface illumination. The term "sensory signals" will be used to characterize the central neural pattern resulting from the sensory processes. The term is not used to imply awareness; nor is it intended to imply that central processes such as figure-ground organization may not enhance or inhibit neural responses (see p. 44). What is proposed is that the sensory signals do not specify a particular percept but allow for alternative percepts. Just as a contour by itself does not specify a right or left edge but allows for reversal of figure and ground (e.g., the well known reversible vase- and faces- Rubin figure, illustrated in Boring, 1942), the central neural encoding of a luminance distribution is assumed to be indeterminate. The word "schema" is used to refer to the representational mechanism that governs the structuring of sensory signals in terms of a specific surface lightness and the deviations due to illumination differences. A schema is a trace representation of a surface color: e.g., a white, gray, or black surface uniformly illuminated; these surface colors shadowed, spotlighted, or in chromatic illumination are encoded in terms of a surface color schema plus a correction for the illumination. The concept of a schema has been used to refer to a classification of past experiences which represents a central tendency, or communality, among stimuli. The empirical origin of schemata has typically been emphasized. As I use the term, the development and utilization of schemata also include organizational processes whose origins may be learned or unlearned. The identification of sensory signals with a perceptual schema involves both inferential and organizational processes. Interpretation of specific cues such as highlights or edge gradients or an inherent organizational tendency to keep surface lightness differences to a minimum may lead to the identification of a stimulus with a perceptual schema that a surface is shadowed. A detailed analysis of the bases for relating sensory signals to a schema can not be given at present. The working hypothesis that I shall adopt is that the assimilation of sensory signals to a schema is such as to achieve a percept as simple as possible consistent with the cue properties of a stimulus pattern.

Although there are many pictures which demonstrate how perception fluctuates between alternative organizations of form, it is difficult to find one which allows for alternative lightness perceptions. (Chapter VIII discusses the limitations of the pictorial representation of light and shade.) The picture taken from Kanizsa (1969) is among the best (Plate 3). Looking at it casually, observers report the stem of the goblet is of approximately constant lightness. Constancy occurs because the stem is seen through a transparent knife and is seen in a reduced illumination. One may, however, look at the picture in such a way that constancy fails and the stripe becomes a gray band on the knife. But when the picture is viewed as a whole, the stem of the goblet is seen through the blade of the knife. The tendency to see the knife as equal in lightness gives rise to the impression of transparency. Past experience with knives can apparently be overcome by the tendency to minimize lightness changes.

The experiments in the air-shadow paradigm suggest that a set for the illumination influences the perception of lightness. According to the hypothesis proposed, this occurs because the spatial, object, and illumination cues cause the surface in shadow to be perceived as a lighter surface in reduced illumination rather than as a darker surface. Woodworth and Schlosberg (1954, p. 446) suggested that the

Plate 3. Alternate lightness perceptions. Is there a gray stripe on the black knife blade, or is it the goblet stem showing through a transparent blade? (From G. Kanizsa, Perception, past experience, and the impossible experiment, *Acta Psychologica,* 31 [1968], 87; by permission of North-Holland Publishing Company, Amsterdam. Photograph courtesy of G. Kanizsa.)

changes in lightness accompanying shifts of attitude in Henneman's (1935) study do not change the apparent luminance, or intensive magnitude, of the stimulus. It would seem reasonable to hypothesize that the changes in lightness in Henneman's experiment are the result of assimilating sensory signals to different schemata. Fluctuations in lighting and the movement of a surface from one position to another have provided specific experiences about the changes in the appearance of white surfaces under different illuminations. What is suggested is that these experiences are stored in terms of a prototypical or normally illuminated white and the corrections arising from the perception of a white in a heightened or reduced illumination. Whether the stimulus is seen as a white in reduced illumination or as a gray depends upon whether stimulus cues and an observer's attitude induce the perceptual system to assimilate the sensory signals to a schema of a white plus a correction for the reduced illumination or to a schema for a gray surface normally illuminated. According to the hypothesis of Woodworth and Schlosberg, the apparent luminance of a color depends only on the intensity information encoded by the sensory signals, and that apparent luminance is preserved when sensory signals are assimilated to alternative schema.

Unlike the hypothesis that an observer discounts the intensity of the illumination in perceiving lightness, the present hypothesis does not assume that there is an automatic central scaling in which perceived lightness is adjusted in terms of the registered illumination. The assimilation of sensory signals to a schema depends upon making contact with specific memory traces. Wallach and O'Connell (1950) found that a shadow cast by a static wire form was seen as three dimensional after observers had been given the specific experience of seeing the same wire figure in rotation. Leeper (1935) has also shown that memory traces resulting from specific past experiences are of great importance in determining which of two alternative organizations will be seen. In addition, the sensory information limits what organizations may occur. Without stimulus support an area of reduced surface luminance will not be seen as a shadow but as a darker surface color, though the observer knows that the reduction in luminance is the result of a shadow (see p. 125). Apparently, the visual cues that normally occur must be present to arouse the pertinent interpretive and organizational processes.

The failure of the factor of illumination to significantly effect lightness judgments in a dark-room setting can be attributed to lack of stimulus support for encoding the sensory signals as a darker (lighter) surface color in a heightened (lowered) illumination. When one views a lighted target in a dark room, one has no impression of a prevailing illumination in the room—only the impression of looking through dark space at a surface and background uniformly lighted by a separate illuminant. Under these conditions, differences in the sensory signals that result from altering the illumination of a target tend to be encoded as differences in surface lightness. Thus, lightness perception is to a large degree consistent with the results of experiments on contrast and adaptation. It is, of course, impossible to be certain that cues to the illumination influence lightness perception in the way suggested. The data strongly suggest, however, that the perception of lightness is not the result of a process which coordinates in an immediate and automatic way the perceived lightness with a reference derived from discounting the intensity of the illumination. The importance of an observer's set, the reported difficulties in making matches, and the large individual differences indicate a looser and more variable perceptual process.

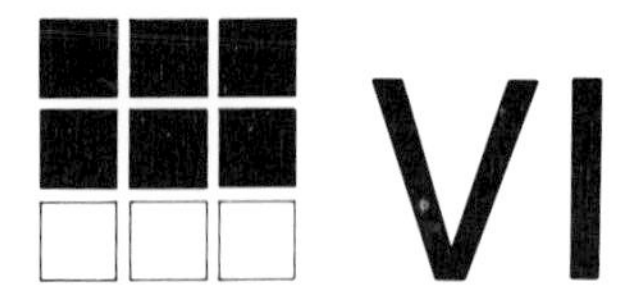

VI

Illumination

The hypothesis that sensory signals are scaled relative to an illumination reference level and the hypothesis that sensory signals are matched to a schema both assume that cues for the illumination affect the perception of color. In this chapter we shall examine how, according to each of these hypotheses, cues to the illumination affect surface color perception.

Helmholtz (*1867*, 1925) was among the first to emphasize the connection between the perception of surface color and the perception of the color of the light by which the surface is illuminated. In the extensive literature that has since developed there are numerous views concerning the relationship between surface color and illumination, two of which we have discussed. According to the first, cues to the illumination are used to establish a reference level with respect to which the color of a surface becomes defined. According to the second, these cues to the illumination may affect the assimilation of sensory signals to a schema by organizational and inferential processes. For example, the *Law of Pragnänz* may include the idea of minimizing the discrepancy between a stimulus and an internal schema. These views differ in their assumptions and implications concerning the relationship between the perceptions of surface color and illumination.

Albedo Hypothesis

The view that an observer discounts the intensity of the illumination in perceiving lightness has come to be called the albedo hypothesis. This hypothesis asserts that processes underlying the perception of lightness produce the same results as those given by the equation $A = S/M$, where M is the registered illumination, S is the intensity of the reflected light, and A the perceived lightness or albedo. A weakly illuminated piece of chalk and a strongly illuminated piece of coal may physically reflect equal amounts of light, but the chalk is seen as white and the coal as black because in each case the illumination has been evaluated and the lightness adjusted. The albedo hypothesis implies that the perception of lightness is the result of combining two items of information: (a) the intensity of the light reflected from a surface and (b) the intensity of the light illuminating the surface. The first is given directly by the light entering the eyes. The second is given by a variety of cues in the field. According to the albedo hypothesis a change in the perceived illumination is a sufficient condition for a change in perceived lightness. A retinal luminance is assumed to determine a unique relation of apparent illumination to apparent lightness. Thus, perceived lightness and perceived illumination are coupled in an invariant relationship. An accurate judgment of illumination implies an accurate judgment of lightness. An enhanced (lowered) impression of the illumination results in the perception of a darker (lighter) surface.

Beck (1959, 1961) has presented evidence that perceived lightness and illumination are not mutually related in the manner implied by the albedo hypothesis. The experiments investigated observers' judgments of the lightness and illumination of a single surface and background uniformly illuminated in a dark room. A binocular matching technique was employed in which the illumination on a standard surface was fixed at different values and observers were given the task of adjusting the illumination on a comparison surface so that it would be equal to that of the standard. Two types of surfaces were investigated. One type consisted of surfaces composed of a clearly perceptible pattern of two different but uniform reflectances and may be described as speckled or striped. The higher reflectances of the standard and comparison surface were the same. A second type consisted of surfaces possessing a surface texture that caused the surfaces to reflect light nonuniformly, for example, a stippled surface produced from shading materials and flannel cloth. There were no clearly discriminable areas of high luminance. The average reflectance of the standard surface was from two to four times that of the comparison surface. These surfaces were presented with and without a white background. Also presented with and without a white background were textureless gray papers. The results of the experiments indicated that judgments of illumination were accurate for the striped and spotted surfaces and were strongly influenced by the higher of the two intensities reflected from the surface. Two surfaces were judged to be equally illuminated when the luminances of the white areas on the two surfaces were approximately equal. For some observers a heightened impression of the illumination occurred with large luminance contrasts. On the other hand, when no clearly discriminable areas of high luminance were present, as on the stippled and wool surfaces, the illumination judgments were strongly influenced by the average luminance reflected to the eyes, and were therefore inaccurate. The black and the gray stippled surfaces were judged to be illuminated equally when the average luminances of the two were approximately equal. When white backgrounds were placed behind the standard and comparison surfaces, the observers' illumination matches became more accurate. The judgments of the observers were then close to what would be expected if they considered the two surfaces equally illuminated when the intensities of the light reflected from the white backgrounds were equal.

At the conclusion of the illumination matches, the illumination of the standard surface was set at the fixed values used in the experiment, and the illumination of the comparison surface was set at the median of each observer's matches. The observer then matched the lightnesses of the standard and of the comparison surfaces to a scale of grays. The results failed to support the implications of the albedo hypothesis that the perceptions of lightness and illumination are coupled in an invariant relationship. For the flannel surfaces, inaccurate judgments of the illumination were accompanied by the complete absence of constancy. For the stippled surfaces, the same inaccurate judgments of the illumination were accompanied by good constancy. For the speckled and striped surfaces, accurate judgments of the illumination were accompanied by good constancy. The results of the experiments suggest that the relations between judgments of lightness and of illumination are based on the properties of the pattern of reflected light. For the stippled surfaces, the presence of sharp luminance differences through internal contrast maintains the lightness of the surface approximately constant when the illumination is changed. With the wool cloths, the lack of sharp luminance differences causes a failure of constancy. However, judgments of illuminations for both the stippled surfaces and the flannel cloths are inaccurate because they are influenced by the average luminance of the reflected light. For the striped and speckled surfaces, on the other hand, both judgments of lightness and illumination are accurate because internal contrast effects again produce approximate lightness constancy, and because the illumination judgments are now based on equating the light reflected from the white areas of the standard and comparison surfaces. The results suggest that the perception of illumination and the perception of lightness have individual stimulus correlates and that the perception of illumination is not a determinant of the perception of lightness but a parallel phenomenon accompanying it. This conclusion is supported by the results of a study of the perceptions of illumination and lightness by Oyama (1968). Beck (1965) and Flock (1970) also report that a change in the apparent illumination is not a sufficient condition for a change in apparent lightness and that apparent lightness does not vary as a linear function of the apparent illumination as implied by the albedo hypothesis (see pp. 131–134). Experiments of Henneman (1935) and Thouless (1931) cast further doubt on the assumption that judgments of lightness and illumination are coupled in a precise manner. Thus, we must either reject the albedo hypothesis or assume that conscious judgments of the intensity of the illumination are not the same as the unconscious taking into account of the intensity of the illumination in determining the perception of lightness.

Lightness and Pronouncedness as Alternative Interpretations

Katz (*1911*, 1935) pointed out that in the surface mode of perception both insistence and lightness are independent dimensions of a surface color. He proposed that insistence

—the intensity of the luminous stimulation (the absolute luminance)—is the basis for the perception of the illumination of a single surface, such as a wall, perceived to be illuminated uniformly. If the surface is not uniformly illuminated, the impression of the overall illumination of the surface is given by the insistence of the dominant parts of the surface, and the impression of illumination of local areas of deviation is given by their insistence. Wallach (1948, 1963) adopted a similar position and suggested that the perception of illumination is closely related to the luminous appearance of a surface (see p. 60). The experiments of Beck (1959, 1961) and Oyama (1968), however, indicate that illumination judgments of a single surface are not a simple function of the absolute luminance of a surface. The judgments of illumination were found to be a function of the pattern of luminances reflected from a field. Illumination judgments were strongly influenced by the maximum luminance reflected when either highlights or clearly discriminable areas of high surface or background luminance were present. The dimension of insistence or brightness may not represent the direct perception of the absolute luminance of a surface as proposed by Katz but a complex attribute, like glossiness, that depends on the pattern of the reflected light. The perception of illumination would not be based on the perception of the absolute luminance of a surface. Rather, brightness judgments of a surface would represent a response to highlights and other nonuniform reflectances from the visual field which indicate the intensity of the illumination.

Though there is experimental support that observers are able to adjust surfaces to look equally bright or equally light, the evidence that the brightness judgment is a direct function of the absolute luminance of a surface is ambiguous. The results only partially confirm that two surfaces are equally bright when they are of the same luminance. Landauer and Rodger (1964) found that observers when instructed to equate the luminance of two surfaces set the comparison surface at a value which was 70 per cent of the luminance of the standard surface. Henneman (1935) performed two experiments in which observers were asked to equate two surfaces for brightness and for lightness. Henneman found that the distinction between brightness and lightness varied with observers and was not made by some observers in a simplified visual field. Henneman reported that the perception of both lightness and brightness became more distinct when the visual field was made more complex. This is consistent with the fact that more cues to the intensity of the illumination are present in a complex field. Thouless (1932) reported that ten out of forty-five observers could not distinguish between lightness and brightness. Of the remaining thirty-five observers, twelve showed only a small difference between a setting in which the comparison and standard surface appeared equal in brightness and a setting in which they appeared equal in lightness, and twenty three showed a large difference in their settings for equal lightness and for equal brightness. In general, observers apparently judged that two surfaces were equally bright when the luminance of the comparison surface was considerably higher than that of the standard.

The best evidence that judgment of brightness is a direct function of the luminance of a surface comes from a recent experiment by Lie (1969a). The targets to be equated consisted of squares placed in front of backgrounds of differing luminances. The luminance of the background of the standard target was set higher than the luminance of the background of the comparison target. In Lie's main experiment, the luminance of the standard target was below that of its background. Lie asked observers to choose the comparison target whose lightness matched most closely the lightness of the standard target. The choice of target proved to be strongly influenced by the relative luminance of a comparison target to its background, which deviated markedly from the absolute luminance of the standard target. Lie then informed the observers that the comparison target, though equal in lightness to the standard, was probably perceived to be dimmer or less bright than the standard. The observers were then asked to disregard differences in lightness and to choose a comparison target that matched the standard target in brightness. The targets that they chose as equal to the standard in brightness were very nearly its equal in luminance.

The procedure of Lie, however, clearly implied that a match of brightness required raising the luminance of the comparison target. It is possible that observers were responding not simply to the luminance of the target but to the contrast between the target and the background. Surfaces which differ greatly in their luminance produce strong edge contrasts that could influence brightness judgments. Moreover, under high illumination the gray scale expands, and large lightness differences could therefore suggest that a surface is strongly illuminated. Beck (1959) found that a heightened impression of illumination of a surface occurs with large lightness differences.

A recent study of Beck (1972) suggests that observers are not able to make accurate judgments of surface luminance when the neural signal is altered by lateral inhibitory interactions. The view that brightness judgments reflect a direct judgment of target luminance is, thus, not supported. Rather, the results are consistent with the hypothesis that observers, in making brightness judgments of a surface, selectively respond to cues for the intensity of the illumination. An implication of the results is that information about luminance is not preserved when a neural signal is altered by lateral inhibition. When lateral inhibition does not modify the neural signal, information about luminance is preserved. Since the results of experiments on lateral inhibition typically show that a field of lower luminance has little if any effect on a field of higher luminance, information about the luminance of a white area or of highlights is preserved. It should also be noted that the results do not say anything explicit about information relating to the average neural excitation produced by the luminances in a visual scene. Models of simultaneous contrast and of constancy have been proposed in which there is assumed a reference level based on the average neural excitation produced by the luminances in a visual scene (Helson, 1943; Marimont, 1962; see pp. 79–80).

What of Katz's dimension of pronouncedness? There is a close relationship between pronouncedness and the impression of illumination. Suppose a white index card is placed so that its face is perpendicular to an overhead light source, and a second white card is placed so that its face is parallel to the same light source. Ordinarily one would describe the two cards as white. However, they do not appear identical. The luminance of the parallel card is less than that of the perpendicular card. This difference in luminance is ordinarily seen as a difference in the illumination of the two cards. The parallel card appears to be white with a gray cast over it. Katz (*1911*, 1935) introduced the term pronouncedness to refer to the fact that the card perpendicular to the light source is a better or whiter white. As in the stem of the goblet (Plate 3), the perceived lightness of the parallel card can vary with an observer's attitude. He may see the reduction in luminance as a difference in the lightness of the two cards, or he may ascribe the gray cast to the surface and see the card as a gray. But accompanying the change in the percept from a white in reduced illumination to a gray, there is an apprehension of an underlying constancy based on the constancy of the sensory signals.

The perceptions of white, gray, and black are not simple sensory discriminations. For a coplanar surface and surround perceived to be illuminated uniformly in a dark room, I have suggested that the sensory signals corresponding to luminance variations would be encoded as differences in surface lightness, enabling lightness perception to be expressed as a function of sensory processes. For a surface perceived to be indirectly illuminated, however, the perception of lightness cannot be specified solely in terms of processes of excitation and inhibition. We must distinguish between a white surface and the goodness or whiteness of a white surface. When both cards are seen as white, the white of the card perpendicular to the light source is a whiter white, or a white of greater pronouncedness than the white of the parallel card, which is perceived to have a gray cast. Differences in pronouncedness appear to result from the fact that the neural correlates of intensity determined by the sensory signals do not always uniquely determine the perception of lightness. The attribute of pronouncedness for achromatic surface colors arises because of an observer's tendency to maintain constancy of lightness. Because of spatial, textural, and illumination cues, there is in many situations a strong presumption to see lightness as the same. Thus, lightness and pronouncedness represent alternative stimulus interpretations. Under ambiguous conditions, an area of reduced surface luminance may be seen, with one attitude, as a white with a gray cast or as a surface of decreased pronouncedness or illumination, and with another attitude as a gray surface color. If every change in the neural intensity correlate of a surface gave rise to a changed perception of lightness there would be no dimension of pronouncedness.

The perceptions of lightness and illumination are related in two ways. First, they may be functions of the same stimulus correlates as in the Gelb effect (see p. 117). The perception of illumination covaries with the perception of lightness, but the perception of illumination is a parallel phenomenon accompanying the perception of lightness rather than a condition for the perception of lightness. Second, as a result of the cue properties of a stimulus and inherent organizational tendencies, processes come into play that determine whether luminance variations are seen as differences in the illumination of a surface or as differences in the lightness of a surface.

This latter relationship, however, is very different from that assumed by the albedo hypothesis. Cues for the illumination represent only one among many factors that may affect the encoding of sensory signals. Thus, an altered impression of the illumination is not a sufficient condition for a change in perceived lightness. The encoding of a sensory signal is a function of multiple factors, such as cues relating to the stimulus pattern (shape, size, and edge gradient of luminance variations, past experience, and organizational factors), as well as the cues for the illumination. For example, the two sides of a book may be seen to be illuminated differently because texture and object cues support the presumption that the cover is of a single lightness. It is not that lightness is evaluated with respect to the illumination, as the albedo hypothesis suggests. Rather, the opposite may be the case. The impression of varying illumination may be the result of a presumption that the surface is of a uniform lightness. The perceptual system, other things being equal, organizes the sensory signals in such a way as to minimize lightness changes.

Experimental results fail to support the hypothesis of a linkage of apparent lightness and apparent illumination. What is proposed is that cues to the illumination can affect perceived lightness under some conditions. Beck (1971) has proposed that these cues affect perceived lightness when they effectively create the impression of a "special" illumination, as in the apprehension of a decreased illumination due to a shadow or an enhanced illumination due to a spotlight. Moreover, the effectiveness of illumination cues in altering perceived lightness apparently depends on highly specific information and not on a general impression of the illumination. A white card with a gray cast may be perceived as a white in reduced illumination because the card is apprehended to be illuminated less strongly. A less strongly illuminated white card cannot, however, be seen as a black in a heightened illumination even when contextual cues suggest that the surface is strongly illuminated (Beck, 1965). To assimilate sensory signals into a schema indicating a brightly illuminated black surface would seem to require surface highlights or other specific stimulus support. Thus, the contextual cues indicating that a surface is strongly illuminated may preclude the perception of a white under reduced illumination, and the surface will then be seen as gray. Since, however, they do not provide sufficient stimulus support for seeing the surface as a black under a heightened illumination, there is no proportional variation between the perceived darkening of the surface and the general impression of a heightened illumination produced.

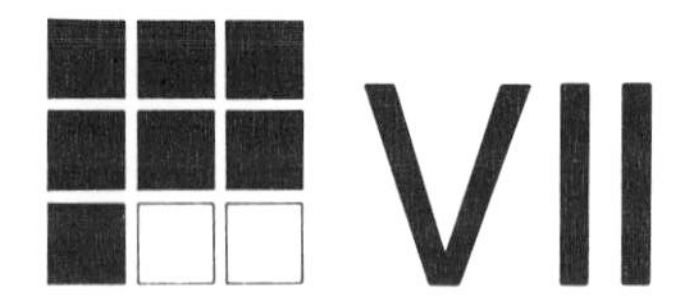

VII

Constancy Experiments, II

The phenomenological study of the modes of color appearance indicated that the perceptions of surface color, illumination, and space are interrelated. The experiments reviewed in this chapter attempt to demonstrate that cues to the illumination and spatial position of a surface affect lightness perception. Although the experiments do not provide a strict test of the relative merits of the albedo hypothesis and the hypothesis of alternative interpretations, they help to clarify the factual background. The topic of memory color is also reviewed in this chapter.

Illumination of a Spatial Region

The experience of illumination in a spatial region is a complex affair involving the integration of many factors such as highlights, shadows, and the contrasts between surfaces strongly and weakly illuminated. These factors combine in ordinary experience to give distinct illumination scenes: daylight, twilight, and night. It has been commonly assumed that if highlights, shadows, and other distinctive cues are eliminated, the impression of illumination is determined by the average color and luminance of the reflected light. Katz (*1911*, 1935) proposed that the impression of the overall illumination is a direct function of the total light flux in the visual field. A high total insistence produces the impression of a strong illumination, a low total insistence the impression of a weak illumination.

Research into whether the impression of the illumination in a spatial region can affect the perception of color has been limited. Experiments have attempted to show that changes in (a) the surface's relative size in the visual field, (b) the actual or apparent spatial position of the surface, (c) the properties of contours delimiting the spatial region in which the surface is located, and (d) the ratios in the luminances reflected from neighboring surfaces in the field alter the impression of illumination of the surface, thereby affecting the perception of its lightness.

Changes in illumination of a scene ordinarily produce changes in the light reflected from a large area of the visual field. Thus, Katz (*1911*, 1935) reported that if a small area of the visual field is viewed through a neutral density filter, a white surface appears gray in the prevailing illumination. If the area of the visual field seen through the filter is increased, the paper appears white in a reduced illumination. Increasing the size of the area of altered luminance concomitantly produces the impression of an independently illuminated region and leaves the lightnesses of surfaces within the region constant. Katz named this effect the *first law of field size*. One cannot, however, conclude that it is the perception of an independently illuminated region that is responsible for constancy of lightness when the change of luminance occurs over a large area of the visual field. Since the central area of reduced luminance is enlarged and the surround area of high luminance is decreased as a larger portion of the field is viewed through the filter, the elimination of the change in lightness may be ascribed to the reduction of the effect of contrast. Katz, however, reported that the same phenomenal change occurs when the retinal stimulus is constant and only the apparent size of a colored area is altered. If an afterimage is projected on a wall, the wall is seen as tinged with the color of the afterimage when the afterimage covers a small part of the wall. According to Katz, the degree of tingeing is a function of the size of the afterimage. When the observer steps away from the wall the afterimage becomes larger. As the afterimage becomes larger the wall is seen as less tinged with the afterimage color and one has the impression instead of a different illumination within the area covered by the afterimage. Katz referred to this case as the *second law of field size,* which asserts that if an afterimage is large it produces the impression of a change in the illumination rather than a change in the color of the surface on which it is projected. Katz's observations, if they are corroborated by further experiments, will show that the impression of a region of altered illumination can change the perception of color.

The illumination of surfaces in approximately the same plane commonly is the same but it differs for surfaces in differing planes. Kardos (1934) has performed many experiments showing how the spatial arrangement of surfaces affects the observer's perceptions of their illumination and lightness. In general, these experiments show that surfaces in the same plane appear to be illuminated equally; those in differing planes may appear to be illuminated differently, and a change in the perceived illumination of a surface is accompanied by a change in its perceived lightness. Figure 7 illustrates the arrangement in one of Kardos' experiments. The hole-screen *HS* was illuminated normally, while the far screen *FS* was shadowed both by the shadow caster *S1* and the left side of the hole-screen *HS*. A disk was successively

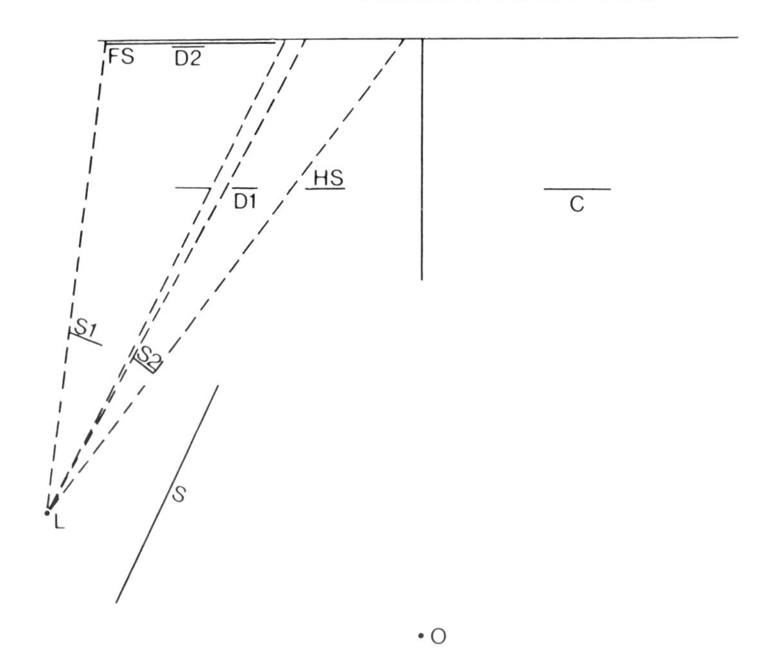

Figure 7. Arrangement in Kardos' experiment: L, light source; S_1, S_2, shadow casters; HS, hole-screen; FS, far screen; D1, disc in near position; D2, disk in far position; S, screen concealing light source and shadow casters; O, observer; C, comparison color wheel.

placed in two positions. In the near position *D1* the disk was located in the same plane as the hole-screen. A shadow caster *S2* cast a shadow on the disk in such a way that the penumbra was not visible. The shadow cast only on the disk was "invisible" to an observer. A screen *S* blocked the observer's view of the light source and the shadow casters. In the far position *D2* the disk was placed directly in front of the far screen. The shadow falling on the far screen fell also on the disk. The luminance of the disk in the two positions was carefully equated and was approximately equal to that of the hole-screen. The disk in the far position was also made slightly larger so that it would subtend the same visual angle as in the near position. The disk was light gray, and the hole-screen and far screen were the same dark gray. An observer seated at *O* judged the lightness of the disk when it was in the plane of the hole-screen and when it was in front of the far screen. The only difference was the perceived location of the disk. Changes in the perceived position of the disk changed its perceived lightness. When located in the plane of the hole-screen, the disk was seen as a strongly illuminated dark gray standing in the same illumination as the hole-screen. When seen through the hole and as standing before the far screen, the disk was seen as a weakly illuminated light gray standing in the same illumination as the far screen. Thus, the disk took on the apparent illumination of the surfaces of the plane in or near which it was situated, with accompanying changes in its perceived lightness. Similar results were reported by Katz (Woodworth & Schlosberg, 1954, p. 444). A white circular disk set up behind an episcotister was viewed by an observer from different distances. When viewed from afar, the disk was seen not as behind the episcotister but rather as a dark gray standing in the overall illumination in the same plane as the episcotister. When viewed from nearby, the disk was seen as a light gray standing in an area of reduced illumination behind the episcotister.

Since substantially the same luminance distribution produced different perceived lightnesses depending upon the apparent position of the surfaces, these experiments indicate that the determinants of perceived lightness include not only the luminance relationships but the stimuli for spatial position. The question posed is, How do stimuli for spatial position influence the perception of lightness? The change from the perception of a dark gray surface color in the prevailing illumination to that of a lighter surface color in a spatial region of lower illumination suggests that the perceived spatial position of the surfaces causes their lightness to change because of the altered impression of the illumination. However, since contrast is decreased when a surface is moved out of the plane of its immediate surround (Gogel & Mershon, 1969), the possibility cannot be completely excluded that variations in contrast account for the lightness changes that occurred.

MacLeod (1932) has shown that casting a shadow on a background surface creates the impression of a separately illuminated region that alters the perceived lightness of a surface standing before it. A light gray disk that stood in a concealed shadow (i.e., the shadow penumbra was not visible) appeared black when no shadow was present on the background. When a visible shadow was cast on the background (i.e., the shadow penumbra was visible), the disk was seen as a lighter gray in shadow. The larger the shadow on the background, the lighter and more shaded the disk appeared. The lightening was greater when the disk was just in front of the background than when the disk was further forward. This effect is consistent with the suggestions of Katz (*1911*, 1935) and Kardos (1934) that the perceived illumination of an area just in front of a background is determined by the background. The lightening of the disk was also enhanced when the background shadow was made tridimensional, i.e., when the disk was seen as standing in front of a shadowed corner. MacLeod tested for contrast effects due to the reduced luminance of the shadowed background by comparing observer's lightness judgments when a background of the same size and luminance as the shadow was substituted for it. All observers reported the disk as lighter when it was standing before the shadow than when it was standing before the equivalent gray background. The lightening of the disk, to be sure, could be ascribed to contrast effects produced by the graded contour or the internal inhomogeneities of the shadowed background, which were not present on the gray background. The fact that contrast effects are minimized when a surface stands in front of a background and when the background is of lower luminance than the target, however, suggests that the shadowed background lightened the disk by making it appear to be in a spatial area of reduced illumination.

Katz (*1911*, 1935) pointed out that the range in luminance changes due to changes in illumination is many times greater than the range in luminance changes due to changes in the reflectance of a surface. The luminance ratio of the whitest white to the deepest black surface is not much greater than 60:1. Differences any greater than this in the luminances reflected from the objects in a scene would, therefore, indicate differences in illumination. Koffka (1935) incorporated Katz's observation into the principles for the organization of a stimulus field. The structure of the perceptual field is such that luminance differences less than 60:1 represent differences in lightness, whereas those that are greater represent illumination differences. Koffka suggested that an area of the visual field which differs from another area by a luminance ratio greater than 60:1 establishes a new illumination reference level in which there is a readjustment in the correspondence between lightness and intensity.

Perhaps the experiment most commonly cited to show that the perceived illumination may affect lightness perception is that of Gelb (1929). In a room illuminated by a weak ceiling lamp, the beam of a projection lantern was focused on a revolving black disk. Only the disk was illuminated by the lantern. Observers reported seeing a white disk standing in the dim general illumination of the room. When a small piece of white paper was held in front of the disk to intercept the beam from the lantern, the percept changed. The disk was now seen as black, the paper as white, and both as strongly illuminated. According to the albedo hypothesis, the disk changed in appearance from a dimly illuminated white to a strongly illuminated black because the white paper had indicated the specific illumination falling on it. The light reflected from the white paper was greater than any that could possibly have been reflected by a surface in the dim illumination of the room. The perceived lightness of the disk was therefore readjusted in terms of the new reference set by the heightened impression of illumination. The experiment, however, cannot be interpreted unambiguously. The change in the experimental conditions that altered perceived lightness also altered the luminances in the field. Changes in perceived lightness may have been the result, therefore, of contrast effects rather than of information about illumination. The introduction of a white paper not only revealed the presence of a hidden light source but also introduced an area of much higher luminance, which on the basis of contrast would be expected to darken the disk. Stewart (1959) showed that the change in lightness of a disk depended on the size of the small white stimulus and its position in the field of vision: the larger the stimulus and the nearer to the center of the field, the darker the disk. Newson (1958) reported similar results. Their findings indicate that the darkening of the disk is at least in part a result of contrast. The change in the perceived illumination of the disk can be accounted for if the perceived illumination is influenced by the maximum intensity reflected from the field in the neighborhood of the disk as shown by Beck (1959, 1961).

ways: black when perceived as illuminated by the spotlight illuminating the back wall, and gray when perceived as an independent area unrelated to the back wall.

In an experimental setup similar to Gelb's, Beck (1971) attempted to distinguish between lightness changes due to cues to the illumination and those due to contrast. A shadow was used to give the impression that a target was strongly illuminated by a spotlight. If changes in lightness are the result of contrast, the presence of a shadow should either lighten the target or produce no change in perceived lightness. If cues to the illumination affect perceived lightness, a shadow that indicates that a target is in an enhanced illumination should darken the target.

Figure 8 shows the experimental arrangement. Two pieces of black matte cardboard were placed on a table in front of white backgrounds. The walls beyond the immediate white backgrounds were flat black. Portions of the two black surfaces and their backgrounds were illuminated by twin projectors that served as spotlights; a slide with a circular ap-

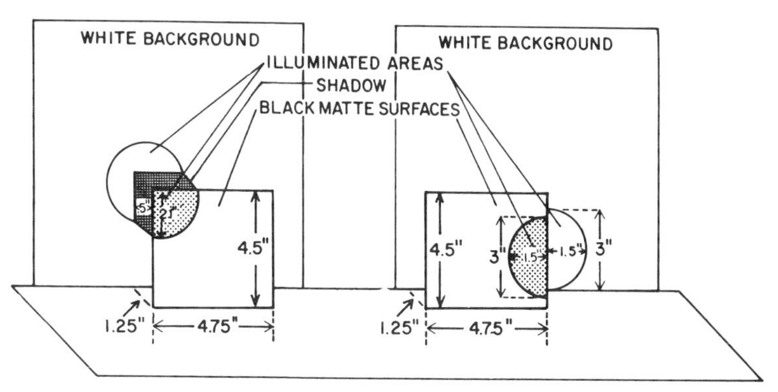

Figure 8. Arrangement in Beck's experiment. (From J. Beck, Surface lightness and cues for the illumination, *Amer. J. Psychol.*, 84 [1971], 3.)

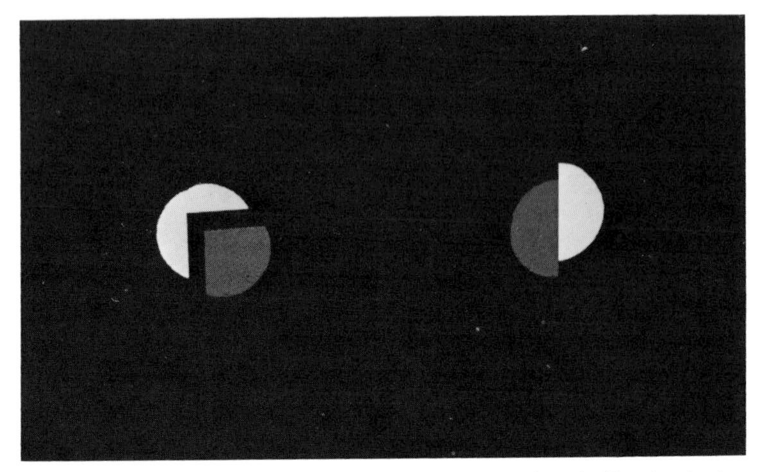

Plate 4. The appearance of a target with a shadow (*left*) and without a shadow (*right*). (From J. Beck, Surface lightness and cues for the illumination, *Amer. J. Psychol.*, 84 [1971], 4; by permission of the University of Illinois Press.)

erture had been inserted into each projector to produce a directed beam of light. The two illuminated areas of the black surfaces served as the targets. Plate 4 shows photographs of the targets, one with the background shadowed and the other with the background unshadowed. In the shadow condition the light fell partly on a corner of the surface and partly on the background above and beside the surface, producing a deep black shadow with sharp contours. In the no-shadow condition the light fell half on the black surface and half on the white background. The illuminated areas of the black surfaces appeared gray. The illuminated areas of the white backgrounds appeared white. The shadow and the unilluminated areas of the black surfaces and white backgrounds appeared black. The luminances of the targets were 14 ft.-L. and of the white backgrounds 180 ft.-L. in both the shadow and no-shadow conditions. The luminance of the shadow in the former ranged between .28 and .5 ft.-L.

After an observer had been familiarized with a nine-step Munsell scale, he was brought into the dark room and stood in front of a screen with a panel set 86 inches from the surfaces. He was told that when the panel was raised he would see two surfaces standing in front of white backgrounds, and was instructed to make a quick and unanalytical judgment of the lightnesses of the two targets, rating them on the lightness scale from 1 to 9. Each of the thirty-nine observers who took part in the experiment made a single rating of the lightness of each of the targets in the shadow and no-shadow conditions. For twenty four of the observers a control condition followed the experimental condition. The viewing panel was closed, and larger pieces of the same black cardboard were substituted for the surfaces. The light from the projectors now fell completely on the pieces of cardboard. The surfaces under these conditions appeared to be self-luminous and could not be matched to a gray scale. In the control condition, the observers were asked to judge if the brightness of the left and right circles of light was the same, or if one circle was brighter than the other. Twenty-nine observers rated the target (the illuminated area of the black surface) in the shadow condition as being darker and ten observers rated the targets in the shadow and no-shadow conditions as having the same lightness. The mean of observers' lightness matches of the target in the shadow condition was a Munsell value of 7.0 and of the target in the no-shadow condition a Munsell value of 8.1. In the control condition, seventeen observers saw the left and right targets as the same brightness, four saw the left target as darker and three saw the right target as darker. The control condition shows that the judged brightnesses of the two targets did not differ when the backgrounds were eliminated.

The fact that a shadow gave rise to an impression of a darker target is inconsistent with a contrast explanation. The shadow, because it is of lower luminance and lies between the black cardboard and the white background, should decrease contrast and cause the target casting a shadow to be seen as either lighter or of the same lightness as the surface without a shadow. The results are consistent with the hypothesis that under some conditions cues to the illumination may affect the perception of lightness. The results suggest that the darkening of a target in experimental set ups similar to Gelb's may be a function of the stimulus information that a target is illuminated by a spotlight, as well as a function of the inhibitory interactions underlying contrast. It apparently depends on the salience of the cues indicating that the target is in a special illumination.

Needless to say, the difference between an observer's judgments of the targets in the shadow and no-shadow conditions is not a simple response to information about the conditions of illumination. The observers realized that the targets were illuminated by projectors. Moreover, the brightness of the white background on the no-shadow side would enable the observers to make judgments of the illumination (Beck, 1961). The equal luminances of the white backgrounds would in fact indicate that the two targets were equally illuminated.

Phenomenally, what the shadow produced was the impression of a surface illuminated by a spotlight. Without the shadow the target was seen as a gray surface and the back-

(1934) that paralleled Gelb's. Kardos cast a shadow on a white disk in such a way that the shadow and the disk coincided. There was no penumbra to indicate a shadow. Observers reported the white disk as a black one, standing in good illumination. However, if the shadow-caster was moved slightly to the side so that the edge of the penumbra became visible, the disk was seen as white with a cast shadow over it. Newhall, Burnham, and Evans (1958) studied how cast shadows affected the perception of hue and saturation as well as the perception of lightness. Three observers, using a colorimeter, matched the lightness, hue, and saturation of ten test colors in light, in a "visible shadow" (the shadow penumbra was visible) and in an "invisible shadow" (the shadow penumbra was concealed). In the shadow-visible conditions, the observers saw a shadow falling on the color sample and part of the surrounding white field as the result of two experimental operations. A large, daylight shadow of medium density (the ratio of shadowed to unshadowed parts of the white surround was 1:5) was cast on a white surround. The luminance of the test color was then reduced in the same proportion by nonselective screens. In the shadow-invisible conditions, no shadow was present on the white surround. The lowering of the luminance was limited to the test color itself. In both conditions, the chromaticity of the color remained unchanged. Considerable constancy of all three attributes occurred. Since small changes in luminance affect hue only slightly, the effect of the shadow on hue was negligible. The matches of lightness and saturation, however, differed under the three conditions of observation. Brunswik ratios were calculated to determine the effects of the shadow on color appearance. The average Brunswik ratio was .74 for lightness and .50 for saturation. The fact that saturation constancy is less than that of lightness constancy is not surprising, since saturation has always been a more subtle and difficult attribute to evaluate.

What aspects of a shadow contribute to constancy? MacLeod (1940) showed that the recognition of a shadow as such does not always contribute to constancy. Two physically equivalent shadows, one encircled to conceal the penumbra, as in Hering's ringed-shadow demonstration, and one unencircled, differed in lightness in the expected way even when neither shadow was recognized as such. The difference persisted when observers were made aware that the darkening of both disks that they perceived was caused by cast shadows. MacLeod (1947) extended these investigations. The contour properties common to shadows were artificially reproduced by means of a rotating disk in such a way as to permit variation in the direction, breadth, and steepness of the artificial penumbrae. He found that a contour gradient approximately equal to that of a shadow penumbra changed the lightness of a gray disk in the direction of a surrounding lighter field even when the graded contour was seen as a gradient and not as a shadow. The change in the lightness of the included area increased with the breadth of the gradient. Steepening the gradient, however, produced the normal contrast effect, i.e., darkened the inlying disk. MacLeod concluded from these experiments that it is not the perception of an altered illumination that is discounted or corrected for in perceiving a surface in shadow as constant in color. That is, shadows need not be recognized as such for constancy. Rather, the important condition for constancy in a shadowed area is the presence of a gradient, or a penumbra. A graded contour changes the lightness of an inlying area in the direction of the surrounding area as illustrated in Plate 5. The photograph shows that a graded contour lightens an area darker than the surround and darkens an area lighter than

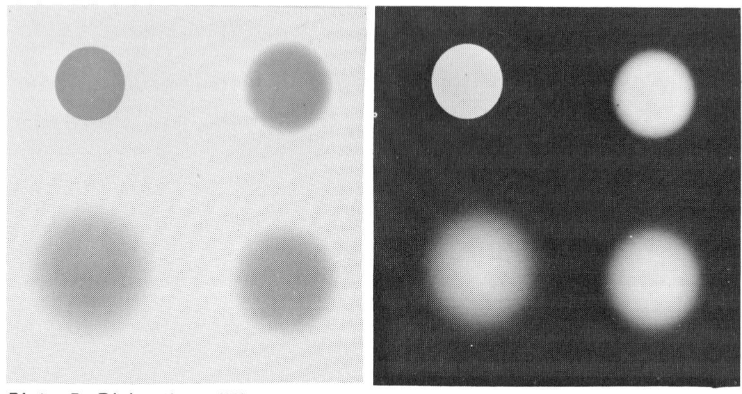

Plate 5. Disks that differ only in the sharpness of their contours. The centers of each disk were of equal reflectance in the original photograph. (From R. M. Evans, *Eye, Film, and Camera in Color Photography* [New York: Wiley, 1959], p. 143; by permission of John Wiley & Sons, Inc. Photograph courtesy of Eastman Kodak.)

the surround. A sharp contour produces the normal contrast effect.

Contour is also of great importance in determining mode of appearance. MacLeod (1947) reported that luminance that changes gradually from a center disk to an outer ring is seen as a transparent membrane or film of color, whereas luminance that changes sharply is seen as an opaque surface color. Koffka and Harrower (1931) also reported that a sharp contour gives rise to a surface appearance. The importance of gradients in determining the mode of appearance was demonstrated by Fry (1931), who found that a black disk on which a white sector was superimposed took on the appearance when rotated of a film, surface, or volume color, depending on the gradient of the white sector. Wallach (1963) has also reported that the impression of a transparent film or surface varies as a function of the nearness of neighboring areas of differing luminance and the "amount" of contour common to these areas. Thus, the relation between surface lightness and the perception of shadow may be expressed as a function of their stimulus correlates. A graded contour has the effect of lightening the inlying field and giving rise to a film appearance; a sharp contour darkens the inlying field and gives rise to a surface appearance.

Contrast effects due to sharpness of contour, however, appear to be not the only factor in the change-over of impressions in the Hering demonstration. Kardos (1934) showed that just making the edge of the shadow visible is enough to change the percept of a black surface to one of a white surface in shadow. Von Fieandt (Brunswik, 1947, p. 124) modified the original Kardos experiment to discover whether observers could be trained to respond to a minimal shadow cue. The shadow-caster was first placed so that a shadow fell only on the disk. No penumbra was visible and the disk appeared black. Moving the shadow-caster closer to the light source so that the penumbra at the edge of the disk became visible caused the disk to appear dramatically whiter. Von Fieandt moved the shadow-caster one hundred times from the position in which no penumbra was visible to one in which there was a visible penumbra. On every twentieth trial, observers matched the lightness of the disk with a minimal shadow cue. The results show that the disk appeared to lighten considerably in the course of the one hundred trials. On the twentieth trial, the Brunswik ratio was .05 and on the hundredth, .61. These results suggest that the recognition of a shadow where minimal cues are present is likely to contribute to lightness constancy. The finding by MacLeod that the conscious recognition of a shadow did not affect the perception of lightness may be limited to the conditions of his experiment, in which there was no visual cue to a shadow. The point has been made that general knowledge without stimulus support is not very effective in altering a percept. Where a shadow is indicated by a minimal penumbra or by object cues such as the continuity of a surface texture, there is a visual basis for separating the shadow from the surface on which it falls. Evans (1948) pointed out that the constancy of a surface in shadow may depend on the attitude of the observer. The attitudes which an observer may take have been described by Rood (Evans, 1948, p. 162) as "looking at," "looking into," and "looking through" a shadow. It would be interesting to determine to what extent shadows without penumbrae may result in constancy when they fall on textured surfaces such as fabrics and wood. If an observer attends to the texture of the surface, the continuity of texture from illuminated to shadowed portions of the surface may be expected to lead to a duo-organization, in which the gray is seen to belong to a film or membrane covering the surface. Lauenstein (1938), moreover, has shown that a penumbra is neither a necessary nor a sufficient condition for the perception of a shadow. Her experiments indicated that the perception of a shadow also depends on the overall organization of the stimulus pattern. No shadow was perceived when the perception of a gray stripe with a graded contour was a good gestalt, i.e., a simple figure.

Object Shadows

An object shadow lies on an object and is created by the orientation of an object to the light source. Object shadows are not seen as shadows but as changes in the orientation and shape of an object's surfaces. A shadowed surface that is perceived as the turning of the surface in the third dimension will exhibit lightness constancy to a considerable extent. But when the position of the surface is seen incorrectly, the surface will be perceived to darken. An example of this

Katona (1935) used the Mach card paradigm to distinguish between the effects of constancy and contrast. A medium gray screen folded at an angle of ninety degrees was placed on a table. The right side of the screen lay in shadow whereas the left side was illuminated. A second screen of the same color, normally illuminated, was placed behind the folded screen. In the center of the shadowed side of the first screen an opening one centimeter square was cut, through which part of the second screen could be seen. The color seen through the hole appeared as a spot on the shadowed side of the folded screen. Experienced observers viewing the stimulus from a distance of one and a half meters could see the stimulus in two ways: They could see the folded screen as it actually was, with the fold toward them and the two sides of the screen extending back; in addition, they could see the screen as lying in a single plane. In the latter case, the observers did not have the impression of the right side in shadow but of two surfaces of different lightness uniformly illuminated. The shadowed right side appeared as a dark gray and the left side as a lighter gray. The observers also saw, on the surface of the dark gray screen, a lighter gray spot. When an observer took an attitude that allowed him to see the real position of the screen, the percept changed. He saw the screen as being of one color, the left an illuminated gray and the right a shadowed one. With this mode of observation, the spot was again seen on the surface of the screen and not as a hole. However, the spot, now seen on a surface of reduced illumination, appeared considerably lighter than when the screen was seen to lie in a single plane. The same results were obtained with inexperienced observers when the apparent spatial position of the screen was altered by varying the distance of observation. The impression of reduced illumination on the right side had a lightening effect on both the background surface and the spot. Analogous results were obtained using a white screen in place of the gray screen. When the shadowed side of the screen was seen to lie in the same plane as the unshadowed, the hole was seen as a spot of darker gray on a light gray screen. When the screen was viewed in its true position, the shadowed side was seen as a shadowed white and the spot was seen to lighten. Katona (1935) argued that the impression of reduced illumination lightens all surfaces standing in it whether they are darker or lighter than the background. This is counter to what would be expected on the basis of contrast, and Katona argued for illumination as a separate factor affecting lightness perception.

Beck (1965) reported two experiments designed to test the albedo hypothesis and to clarify the relation between the perceptions of lightness and illumination. The first experiment used the Mach card figure. A folded card was placed on a table behind a cardboard partition so that its left side was illuminated and its right shadowed. Apertures cut in the screen fixed to one end of the table allowed observers to see the target with either monocular or binocular vision. Observers were used who were unfamiliar with the Mach card demonstration. Changes in the apparent position of the target were induced by drawing rows of dots making an angle of thirteen degrees with the horizontal on each side of the target. When cues to slant were eliminated by restricting an observer to monocular vision, the tendency to see the dots arranged in straight lines caused the target to appear to stand flat and uncreased, directly behind the partition. In the first of two conditions the surfaces were illuminated by an overhead ceiling light. When the target was apparently flat, its sides of lower and higher luminance were seen to be illuminated equally. In a second condition, a projector located below and in front of the target but hidden from the observer was added to the overhead light. The projector beam actually missed the target, falling only on the cardboard partition and the wall behind. However, the target when perceived to stand flat directly behind the partition appeared to be illuminated by the bright light of the projector. When the target was viewed binocularly, it was seen in its correct spatial position and illumination.

Observer's matches of the lightness of the shaded side of the Mach card were slightly darker when the target was seen as apparently flat than when the target was seen in its true spatial position. The small changes in lightness that the observers noted, however, are not consistent with the albedo hypothesis. According to this hypothesis, for a fixed luminance there is a proportional variation between changes in apparent illumination and lightness. Had the target actually been located in the perceived position, the illumination of the shadowed side of the target would have been increased by a factor of 1.5 in the first condition and 58.9 in the second condition. The decrease in lightness matches made by observers when the target was seen as flat and when seen in its correct position was .5 Munsell step in the first condition and 1.25 Munsell steps in the second condition. Thus, the small darkening of the shadowed side when the target is seen as flat is inconsistent with the albedo hypothesis. Similar results were obtained with trained observers who were able to change the apparent position of the card without changing from monocular to binocular observation.

The second experiment used the experimental arrangement of Hochberg and Beck (1954). They placed a trapezoid so that it could be seen either as a trapezoid standing upright or as a square lying flat on the table. Physically the trapezoid stood upright. Many cues were given to the direction of the objective illumination. When the illumination came from overhead, the target appeared lighter when seen as a trapezoid upright (illumination parallel to the perceived target surface) than when seen as a square lying flat on the table (illumination perpendicular to the perceived target surface). When the illumination came from in front at a height the same as the target, the target appeared lighter when seen as a square lying flat on the table (illumination parallel to the perceived target surface) than when seen as a trapezoid upright (illumination perpendicular to the perceived target surface). When the illumination came from the side at a height the same as that of the target, no change in lightness was perceived: the target appeared to be of the same lightness whether seen as a trapezoid standing upright or as a rectangle flat on the table. In this condition the illumination was parallel to the target surface in both of the two perceived positions. Thus, a perceived change in target position with respect to the direction of illumination caused a change in perceived lightness. A fixed target under constant illumination was judged darker when made to appear perpendicular to the direction of illumination than when made to appear parallel to the direction of illumination.

Informal observations have indicated that the actual position of a target is important in determining whether an observer's lightness judgments are altered with changes in the target's apparent position. Beck (1965) examined the effect of target position. Ample cues were present to indicate the direction and distribution of the illumination, which came from overhead. The target, an upright trapezoid, was placed in four positions on a table, in each of which the degree of shadowing of the target was different. The amount of change in apparent illumination resulting from changes in the apparent position of the target also varied. The change of the target from apparently upright to apparently horizontal altered the perceived illumination most strongly in position 1. The illuminance of the central area of the target was 13.5 ft.-c.; the illuminance of the central area of a gray paper of the same reflectance as the target placed flat on the table to correspond to the target projection was 140.6 ft.-c. The illuminances of the target and of a gray paper of the same reflectance as the target placed to correspond to the target projection on the table were 31.3 ft.-c. and 125 ft.-c., respectively, in position 2; 56.3 ft.-c. and 93.8 ft.-c., respectively, in position 3; and 62.5 ft.-c. in position 4. Thus, if the target were actually a rectangle flat on the table, the illuminance of the surface would be increased by a factor of 10.4 in position 1, a factor of 4 in position 2, a factor of 1.7 in position 3, and would not change in position 4.

A significant lightness change occurred only in position 1, though in positions 2 and 3 the many cues to the illumination would indicate the target to be more strongly illuminated when apparently flat. The median darkening in position 1 when the target was apparently flat was only .5 Munsell step. The results again are inconsistent with the albedo hypothesis. The many cues to the illumination would be expected to indicate that the target in positions 2 and 3 would be seen as more strongly illuminated when seen as flat than when seen as upright. The results show that a change in apparent illumination is not a sufficient condition for a change in apparent lightness. The small changes in lightness that occurred in position 1 showed again that apparent lightness does not vary as a proportional function of the apparent illumination, as implied by the albedo hypothesis. Flock (1970) has reported similar results.

The results of the Mach card experiments and rectangle-trapezoid experiments indicate that lightness perception may be affected by how an observer perceives a surface to be oriented with respect to the illumination. However, as I have mentioned, the small changes of lightness and the fact that lightness fails to change under some conditions are inconsistent with the hypothesis that the apparent position of a surface relative to the illumination is used as a basis for computing the reflectances of a surface (i.e., the albedo hypothesis). Rather, the studies seem to support the general hypothesis that the lightness changes that occur are the result of processes of perceptual organization that come into play in dealing with the distributional properties of reflected light. These processes affect whether a variation in luminance will be seen as a difference in the illumination of a surface or in its lightness. Beck (1965) has interpreted the results as supporting the hypothesis that the cue properties of stimuli affect lightness perception by affecting the way in which sensory signals are assimilated to a schema. He has suggested that when a surface reflects a distribution of luminances, its perceived lightness depends upon a decision as to whether or not the surface is of a uniform lightness. If the cue properties of stimuli lead to the decision that a surface is of a uniform lightness, a single lightness is fitted to the varying luminance pattern, and deviations are perceived as differences in the orientation and illumination of the surface. Factors favoring the presumption that a surface is of a single lightness favor constancy. Katona (1929) reports that object character, movement, and continuity of contours all favor constancy. The perceptual system may also prefer a particular organization; it may function to minimize lightness changes if conditions are consistent with variations in luminance being seen as changes in illumination. Thus, when the target is seen as folded, the shaded side of the target is seen as indirectly illuminated, and the area of lower luminance is

The change of lightness in the Hochberg-Beck (1954) paradigm indicates that, although limiting the subsequent integrations that may occur, a particular sensory pattern may be indeterminate. Sensory signals can evoke a perception of a lighter surface with shadows when the surface is seen as turned away from the light and as a darker surface with light spots when the surface is seen as facing the light. What is seen as a shadow, a light spot, and a normally illuminated surface is made consistent in each case with the apparent position of the surface. In general, whether a change in the apparent position of a target induces consistent changes in surface lightness may be expected to depend on the pattern of luminance variations over the surface. The consistently darker judgments when the target was apparently flat in position 1 may reflect the fact that the target in this position had been seen to be almost entirely shadowed when seen upright. Since the illumination came from overhead, the perception of a heavily shadowed surface was inconsistent with seeing the surface as flat and therefore tended to induce a new encoding of the sensory signals.

Beck (1969) has also reported an experiment which indicates that the orientation of a photograph with respect to an observer's eyes affects the perception of lightness. Observers were shown, first right side up and then inverted, a photograph of a young girl dressed in her bathing suit (see Plate 6). They reported that the chest of the child seemed to be a lighter gray when they looked at the photograph right side up than when the photograph was inverted. After the observers had made their lightness judgments, they were asked whether the space between the child's chin and chest appeared equal in the two positions of the photograph. They reported that the space looked greater when the photograph was upright.

Plate 6. Orientation and lightness perception: a photograph. (From R. M. Evans, *Eye, Film, and Camera in Color Photography* [New York: Wiley, 1959], p. 279; by permission of John Wiley & Sons, Inc. Reproduced in two orientations, from J. Beck, *Amer. J. Psychol.,* 82 [1969], 369; by permission of the University of Illinois Press.)

Careful observation of Plate 6 suggests an important difference between the upright and inverted positions. In the upright position the photograph of the child looks normal. When the photograph is turned upside down there is a disorganization of the percept. The pattern of light and shade on the face, for example, is no longer seen as belonging to one surface but tends to be seen as separate, unrelated areas. Many observers also reported that the dark area at the base of the child's neck appeared much clearer when the photograph was inverted with respect to their eyes than when it was upright. In the upright position this black band was indistinct, and the observer had to look for it; when the photograph was inverted the black band was prominent and emerged at once. What is suggested is that when the photograph is upright we tend to see the chest and face of the child as single surfaces of a specific lightness. Variations of luminance are then seen as shadows and highlights that show the modeling of the surfaces. For example, one sees areas of the child's face that are illuminated and areas that are shadowed because of differences in the orientation of these areas to the light source. When the photograph is inverted with respect to the observer's eyes the relatedness of the areas changes. The photograph is seen as made up of separate areas and the image appears flatter. When looked at as separate areas, the luminance variations are seen as differences in lightness. The figural organization of the pattern may also exert a selective effect that emphasizes some part or aspect of the percept. Thus, the band at the base of the neck is clearer when the photograph is seen as a surface made up of multiple lightnesses than when it is seen as a surface with a single lightness.

The difference between observer's judgments of the photograph in the upright and inverted position is not a simple response to the information in the photograph. The inverted photograph is clearly a picture of a child, and the most likely hypothesis of the darkened area is that it is a shadow due to the orientation of the child's body with respect to a light source. Perception, however, depends on highly specific information such as the orientation of the photograph with respect to the eye. When the photograph is upright, the lightness variations apparently evoke a schema that leads the observer to see the areas of the photograph as surfaces of a single lightness varying in orientation and illumination. When the photograph is inverted, this schema is not evoked, and the areas are seen as separate and differing in reflectance.

Transparent Colors

Helmholtz (*1867*, 1925) reported observations showing that it is possible to have a simultaneous perception of two colors, one behind the other. A simple example is the reflection of a white card in a black tile. The white card is seen behind the black tile, and the impression produced is similar to that of a white card in a reduced illumination. If the reflection of the white card is localized in the same plane as the black tile and not behind it, the reflection may give the appearance of a gray spot on the black tile. Fuchs (1923) investigated the conditions under which colors are seen one behind the other. According to him, transparency is not possible if two rectangles of light, which are the same size but of different colors, are superimposed. Under these conditions one sees only a surface whose color is a mixture of the two colors. Transparency may occur when a smaller rectangle is seen behind a larger rectangle. Transparency occurred most readily when parts of the figure in front extended beyond the borders of the figure behind it. Even when stimulus conditions favor the appearance of transparency, however, it is lost if an observer adopts a critical attitude in which he concentrates upon individual points or contours of the surface and does not apprehend the surfaces as independent whole figures.

Heider (1932) reported that the impression of transparency depended on organizational factors favoring the apprehension of two superimposed surfaces. Heider placed a blue episcotister in front of a number of small yellow disks arranged to form a circle on a black background. The upper half of the circle of disks was visible over the top of the episcotister. The lower half of the circle of disks was visible through the blue episcotister. She found that the continuity of the pattern was an important factor in producing the phenomena of transparency. When the upper half of the circle of disks was hidden and an observer saw only the lower half of the circle, the disks appeared a neutral gray. When the upper half of the circle was exposed, an observer saw the yellow disks through a transparent blue film. Prentice, Krimsky, and Barker (1951) reported the results of an experiment attempting to repeat Fuchs' and Heider's observations. Observers judged the apparent yellowness of a yellow ring located (a) in the same plane as a blue episcotister, (b) four feet behind a blue transparent film produced by rotating the episcotister, (c) when half of the ring was covered by the blue film and half was not, and (d) same as (c), but with the two yellow arcs of the circle curved in the same way. According to the results of Fuchs and Heider, the ring should be yellowest in conditions (b) and (c), where good transparency conditions existed and the observers saw the ring four feet behind the blue film produced by the episcotister. The ring should appear least yellow in condition (a), in which the ring and episcotister appear in the same plane. The portion of the ring covered by the blue film should be seen as yellower in condition (c) than in condition (d), where the two arcs did not constitute a simple symmetric figure. Prentice et al. failed to find that the yellow in the ring was enhanced by either the separation in depth of ring and film, the extension of the contours of the ring beyond those of the film, or by the symmetry of the figure. The number of degrees of blue and yellow required to match the yellow ring remained the same when the yellow ring was placed four feet behind the episcotister, when only half of the ring was obscured by the episcotister, and

pattern. The perception of transparency may also depend on the ability of an observer to visualize what the surface would look like through the color of the film. Without specific past experience the phenomenon of transparency may be difficult to obtain. The appearance of a yellow through a blue film is not as common as the appearance of white through a dark film. The perception of transparency appears to involve perceiving a surface color in a particular lighting. White that is seen through a dark film looks similar to white that is seen when the illumination is reduced, as in the evening. Thus, observers have had specific experiences with the appearance of a white color in reduced illumination that favor the perception of transparency.

The frontispiece illustrates the perception of transparent colors. The area of intersection of each pair of circles is the result of combining paints of the two colors in equal amounts. In each, the area of intersection can be seen as either the mixture color when both circles are seen in the same plane, or as the color of one circle seen through the transparent color of the other circle when the circles are seen in different planes. For example, in the pair at the top the area of intersection can be seen as a yellow-green, a green behind a transparent yellow or a yellow behind a transparent green. The transparency effect produced by the yellow and blue circles in the pair at the bottom follows the law of subtractive color mixture rather than that of additive color mixture.[5] Plate 3 shows that the perception of transparency,

[5] When we place a yellow glass in front of a blue surface, subtractive color mixture occurs. The light stimulating the eyes is that which remains following the selective absorption by the blue surface and the yellow glass. Thus, the eyes are stimulated by predominantly green light. When we place a yellow episcotister in front of a blue surface,

however, is not in accordance with our past experience with specific objects. Transparent knife blades are not commonly seen. In fact, the perception would be expected in accordance with a Prägnanz principle, the perception of transparency being simpler in terms of the number of different shapes and lightness values involved than the alternative of seeing a blade with a gray stripe hiding the goblet stem. Kanizsa (1969) has convincingly shown that empirical factors that may influence transparency depend on the contour, shape, and the color of stimuli and not on the meanings of objects. Recognizing a stimulus as a particular object appears to be separable from seeing the shape and surface properties of the object (Gibson, 1950).

Judd (1960, p. 79) has suggested that an orderly pattern of color variations produces the impression of transparency because the light may be perceived as coming from two distinct

---

additive color mixture occurs. The eye is stimulated by the yellow light reflected from the episcotister and the blue light reflected from the blue surface. The physiological effects of these two conditions are quite different. Though the physical situation always differs, the physiological effects resulting from additive and subtractive color mixtures may be similar. Looking at a green surface through a yellow glass causes the eyes to be stimulated by spectral yellow-green light. Looking at a green surface through a yellow episcotister causes the eyes to be stimulated by yellow light and green light. The cone types stimulated by spectral yellow-green light and by yellow and green light are the same. Subtractive color mixture is what predominates when we look at colors through tinted glass, or when paints of different colors are mixed. If I am correct in suggesting that specific experience plays a role in achieving the transparency effect, the phenomenon of transparency might occur more readily with subtractive than with additive color mixtures. The transparency effect to my knowledge has not been investigated as a function of whether subtractive or additive color mixture occurs.

sources. A stimulus consisting of a constant chromatic component superimposed on a varying achromatic component is seen as a constant red film overlying differing shades of gray because the brain interprets a stimulus pattern in terms of the best guess as to what is objectively there. Hering (Fuchs, 1923) argued that depth and figural cues are not sufficient to separate a color into two overlapping figures if the two colors are perfectly uniform and neither possesses any visible texture. Although Hering was apparently mistaken, it is undoubtedly true that differences in texture greatly facilitate the impression of transparency. The presence of two textures ordinarily compels the perception of two overlapping figures.

Memory Color

Hering (*1874*, 1964) has suggested that the characteristic color of an object becomes attached to an object and is an important factor in constancy. A familiar object tends to be perceived in accordance with its daylight color even though the light reflected from the object differs considerably from that reflected under daylight illumination. We expect to see grass as green, bananas as yellow, and the sky as blue.

Duncker (1937) reports one of the first experiments designed to investigate the influence of memory color on color perception. On white paper, he pasted a leaf and a donkey both cut out of green paper. The leaf and donkey were alternately lighted with a red illuminant. A circle cut out of the same green paper appeared gray with only a touch of green when exposed to the red illumination. Observers matched the color of the leaf, donkey, and circle with a color wheel. Duncker found that the mixture on the color wheel required 60-degrees green to match the leaf and 29-degrees green to match the donkey (the average for his six observers). It is interesting to note that for three of Duncker's observers no difference was obtained between their matches of the leaf and the donkey. Duncker suggests that the failure of memory color in these cases was due to each observer's taking a critical attitude in which the color was abstracted from its objective connotations. The effects of memory color are most strongly exhibited when an observer adopts a perceptual attitude directed toward the true color of an object. The donkey and leaf were not exposed simultaneously, since this would have made the observer more likely to adopt a critical or abstract attitude. Duncker also proposed that memory colors, to be effective, must have some degree of stimulus support. That is, memory color may increase the perceived amount of a color rather than create a color that has no stimulus basis.[6] Bruner, Postman, and Rodrigues (1951) obtained similar results. They found that normally red objects (a tomato, a boiled lobster claw) were judged redder, and normally yellow objects (a lemon, a banana) were judged yellower than neutral figures (an oval and an ellipse). The stimuli were cut from gray paper, and the colors of the figure were the result of contrast due to placement on a blue-green background. The stimuli and the matching color wheel were separated by eighty degrees, and observers made their matches by successive comparison. No effect of memory color was found when, in place of the induced color, the stimuli

[6] The results of an experiment by Bruner and Postman (1949) suggest that this may not always be the case. Observers who had been presented, in brief exposures, with playing cards in which normal color and suit associations were reversed, e.g., a black four of hearts, gave color reports in conformity with past experience, e.g., a red four of hearts. However, the instructions to the observer to report everything he "saw or thought he saw," may have facilitated guessing.

were cut from a well-saturated orange paper. Bruner, Postman, and Rodrigues (1951) found an interaction between the variable of memory color and the conditions of stimulation. Memory color is effective in influencing color perception when the color information is poor or unstable. Memory color does not have an effect when the stimulus information is clear cut, as when the figures were cut from orange paper.

Duncker (1939) and Bruner et al. (1951) have shown that under impoverished or marginal conditions the apparent color of an object may change in the direction of its characteristic color. Bolles, Hulicka, and Hanly (1959) criticized both Duncker and Bruner et al. because an observer in these experiments could not make an exact color match. Bolles et al. repeated experiments similar to those of Duncker and Bruner et al. but introduced modifications that allowed observers to make exact color matches. Under these conditions, there was no influence of memory color. There was no difference in the matches made of the donkey and leaf figures. On the basis of these results Bolles et al. (1959) asserted that memory color effects are the result of response biases when an exact equation of the colors is impossible. When an exact color match is possible, an observer will make it despite the possible influences of memory color.

Harper (1953) examined the influence of memory color under good conditions of observation. To eliminate the ambiguity of earlier procedures, he employed the indistinguishability of a figure from its background as a measure of the perceived color. Using orange paper, Harper cut characteristically red figures (an apple, a heart, and a lobster), and figures which had no characteristic color (an oval, a triangle, and the letter Y). The figures were placed in front of a red background, whose degree of redness was varied. The observer's task was to report, as the color of the background was varied from red to orange, when the figure was no longer visible. Harper found that figures having no characteristic color required less red in the background before they became indistinguishable from the background than figures having a characteristic red color required. The number of degrees of red at which the figures became indistinguishable from the background was 71.3 for the noncharacteristically red figures and 134.9 for the characteristically red figures. Since a reasonable assumption is that the criterion of indistinguishability implies that the figure and background were seen to have the same color, the results would appear to indicate that memory color affects perception and does not affect matches because of response biases. Fisher, Hull, and Holtz (1956) repeated Harper's experiment and obtained much smaller differences between observer's matches of neutral and red-associated objects. Fisher et al. also studied the effects of variations in the area, form, contour, and complexity of the figures used. They found that all these variables were apparently involved and could affect the perception of color. In demonstrating the effects of figural properties, Fisher et al. questioned whether the differences obtained by Harper were not the result of figural properties rather than the connotations of the figures. Delk and Fillenbaum (1965) also employed Harper's procedure and found differences in observers' matches depending on whether the figures were red associated (heart, apple, lips), neutral associated (oval, circle, ellipse), or associated with another color (horse, bell, mushroom). The red-associated figures required significantly more red to be mixed with the background color than did the other figures before the observers judged that the figures and the background were the same color. Differential in-

tions may influence color perception, although in what way is still unknown. It is possible that the effects of memory color are brought to bear by means of images. Perky (1910) showed that images may be mistaken for percepts. The confusion of percepts and images would allow color perceptions to be modified by image colors. Another possibility is that the perception of part of a stimulus is sufficient for seeing the entire stimulus. Repetition, for example, enables an observer to see more with a brief exposure (Haber, 1965), suggesting that memory color may be based on using a part of a stimulus for the whole. There is a tendency to fill in a familiar stimulus instead of examining it completely. This appears to be the basis for the overlooking of misprints in reading a familiar text, and would account for the fact that memory color does appear to require sensory support. It would in addition be consistent with the finding that memory colors are most effective when the stimulus information is impoverished. Because we scan selectively, it may also be that identification causes the eye to skip over discrepant color information and see the familiar color of an object. Memory color effects are least when an observer examines a color patch carefully, noting each variation of color in the stimulus. The question usually raised is whether the effect of memory color is a true perceptual effect or a response bias. When response bias is interpreted in terms of the class of responses available to an observer, the distinction is clear; for example, the criticism by Bolles et al. (1959) of memory color experiments in which an exact color match was not possible (see p. 147). When the question, however, is interpreted as referring to whether memory color effects depend on selective scanning but disappear under conditions of careful scrutiny and observation, the distinction between a perceptual effect and a response effect becomes subtle. The color revealed by careful scrutiny may simply reflect differences in the operation of perceptual mechanisms when an observer attends to all the features of a stimulus.

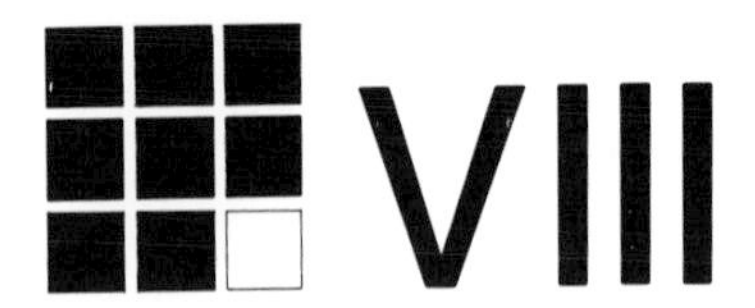

VIII

Surface Textures and Qualities

How a color looks is closely tied to the material on which it appears. Because of this close connection, there are many different industrial color scales (Judd & Wyszecki, 1963). Wright (1967) presents an excellent description of the subtle relations between color, spatial, and textural properties. As pointed out, the colors that are tied to a specific material appearance include those of metals such as gold, silver, copper, and so forth. There is no single specifiable physical property that produces the perception of silver. Silver appears to involve the integration of a complex light-dark reflectance pattern. Characteristically, there are two important aspects that produce the impression of silver, though whether they are necessary conditions remains for experimentation to decide. First, the impression of silver arises when the lightness pattern involves a gradual change from white to gray to dark gray. These local changes in lightness may be attended to and seen as distinct lightness values of the surface. Second, these lightness gradations appear not on the surface, but somewhat indefinitely below it. The top surface appears as a clear film overlying the lightness variations. Phenomenologically, silver exhibits a whole character in the same sense that a square and a circle do. The observer does not ordinarily attend to individual discriminable features and responds to the entire pattern as an integrated whole, the quality we call silver.

From a physical point of view, the different surface color appearances are related to the complex ways in which surfaces may reflect light. For example, the textures of an apple and an orange determine the gradients of intensity and wavelength that produce the characteristic smooth, glossy appearance of the apple surface and the more matte, rough appearance of the orange surface. Similarly, the difference in the appearance between an unpolished and a highly polished wood is due to the difference in the amounts of diffuse and specular light reflected and to the spectral distribution of the light reflected. The deeper color produced when a wood is polished is the result of multiple reflections. Some of the light reflected from the bottom layer is not transmitted through the top layer of polish but reflected back to the bottom surface, where selective absorption takes place for a second time. Polishing the wood changes both the amounts of specular and diffuse reflection and the spectral composition of the light reflected. Polishing also tends to make the top surface invisible since the polished top surface does not possess irregularities upon which an observer may fixate. He tends to see the grain or texture of the surface appearing below the top surface. Apparently an observer is sensititive to the fact that the specular top component of the light reflected is in front of the plane of fixation.

The visual system, it seems, is sensitive to a variety of modifications in the light pattern produced by physical surfaces. These have been studied only to a limited extent, however. Transparency, such as that perceived in panes of glass or in a clear liquid, appears to be correlated with the degree to which light is transmitted regularly. If all the light passed regularly through a pane of glass, the glass would be invisible. For the pane of glass to be seen, some of the light must be reflected from scratches, dirt marks, or pimples. A continuous pane of glass or layer of polish is seen when the stimulus consists of inhomogeneities which reflect only at occasional points.

Glossiness

The surface quality most extensively investigated is glossiness. Hunter (1937) has distinguished five kinds of glossiness. Table 2 lists them and their correlates in terms of the pattern of light reflected to the eye. Glossiness, in general, is correlated with specular reflectance. The different types of glossiness listed in Table 2 correspond to different patterns of specular and diffuse reflectances. Sheen, for example, occurs when a surface reflects specularly the incident light that falls upon it at a grazing angle. When the incident light is perpendicular to the surface, the surface may not appear shiny. However, some surfaces exhibit the opposite characteristics; they appear shiny when the light falls on them at almost any angle except a grazing one (Hunter, 1937). Surfaces that form distinct images also appear glossy. Plate 7 shows how highlights produce the impression of a glossy surface. The highlights make the entire surface appear glossy. When the highlights are taken out, there is a change in the appearance of the entire surface. The appearance of glossiness may also be furthered by the fact that the image appears below the surface. This may introduce a stereoscopic effect and a rivalry between the fixating of the surface texture and the fixating of the images reflected in the surface. Another type of glossiness is lustre, which has been the subject of much debate (Bixby, 1928). It seems generally agreed that a

Table 2. Various kinds of glossiness and the functions of directional reflectance with which they correlate

Kind of glossiness	Function of directional reflectance	Diagram of the angular conditions
Specular	Ratio of $R_{60,-60}$ for the specimen to that of a perfect mirror	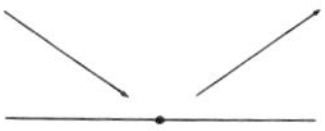
Sheen	Ratio of $R_{85,-85}$ for the specimen to that of a perfect mirror	
Contrast	$R_{60,-60}/R_{60,0}$	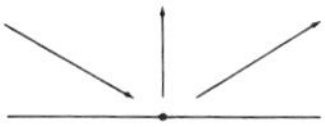
Distinctness of image	Rate of change of $R_{i,-\theta}$ with the angle of incidence i, where the angle of view, $-\theta$, differs by a few minutes of arc from $-i$, that of mirror reflection	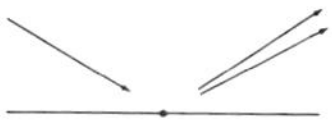
Absence of bloom	Ratio of $R_{i,-i}$ to $R_{i,-\theta}$, where the angle of view, $-\theta$, differs from the angle of mirror reflection, $-i$, by a few degrees	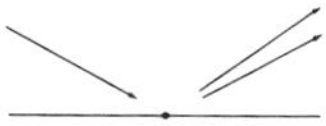

Reproduced from D. B. Judd and G. Wyszecki, *Color in Business, Science, and Industry* (New York: Wiley, 1963), p. 375. Copyright 1963. By permission of John Wiley & Sons, Inc.

surface has lustre when its reflection characteristics change sufficiently with angle of view so that each eye of an observer receives different intensities or qualities of light.

Beck (1964), investigating the relation between the perceptions of lightness and glossiness, found that a glossy surface tends to exhibit greater constancy than a matte surface. The difference between these surfaces is related to the fact that the glossy surface reflects light nonuniformly whereas

Plate 7. Highlights and glossiness. A surface with highlights appears glossy; withc them, it appears matte.

the matte reflects light uniformly. Beck proposed that the difference in the perceived lightness of a matte and a glossy surface can be analyzed in terms of a two stage process. First, contrast processes act to determine a sensory pattern in accordance with the distribution of luminances reflected. The luminance variations of a glossy surface introduce internal contrasts which account in part for the greater constancy of a glossy surface compared to a matte surface when the illumination is changed. Second, as a result of the cue properties of the stimulus, the sensory pattern is organized as a single lightness, with deviations seen as highlights and light spots. What is suggested is that, because of the shape, size, edge gradients, and distribution of the luminance variations, areas of luminance of a surface differing from a particular luminance are seen as light spots and highlights rather than as areas of different surface lightness. Glossiness is not a simple attribute, and its perception involves the organization of a lightness pattern in terms of a specific surface lightness and deviations due to highlights. The perceived lightness of a glossy surface is determined, not by the average luminance reflected, but by a particular luminance within the array of the luminances reflected.

The general problem raised is: What are the means by which a structured pattern is achieved? A surface does not reflect light uniformly; it reflects a distribution of luminances. This distribution, however, yields a specific perception of surface color, texture, and illumination. For example, consider a gray scarf. Typically, one is able to identify (a) the wool material out of which the scarf is made, (b) the pattern of knitting that determines the texture of the scarf, (c) the lightness and color of the scarf, (d) the overall illumination falling on the scarf and the distribution of light and shade, and (e) the folds and creases of the scarf determined by the shape in which it has fallen. These aspects are tied together in an organized structure and are separable from the perception of the identity of the object as a scarf. Processes of figure-ground and unit formation appear to be involved in integrating the sensory information into distinct units and structural relationships that correspond to the different properties of a surface. Gibson (1950) also has argued effectively that much of the structure of a percept has its basis in the stimulus gradients present in the light reflected by an object. For example, Gibson would assert that the features present in the perception of the wool scarf correspond to information in the pattern of the reflected light. There are hierarchies in the variations of luminance that correspond to the structure of the percept. There are, first, variations in the luminances of the wool fibers that make up a wool yarn. The absolute intensities of the fibers vary with the illumination of the yarn. The ratios of the luminances reflected from different wool fibers, however, are the same. Similarly, the absolute intensities of the wool yarn making up purl and knit stitches may vary depending on the illumination. The ratios of intensities again, however, would remain constant. The same would be true for larger areas of the wool scarf. The absolute luminances reflected from different patterns of stitches will vary according to the illumination; the ratios of luminances will again remain constant. Thus, there are patterns of intensities within patterns of intensities which could provide the basic information for the structuring of the percept. We can not critically consider Gibson's assumption that all the information for a percept is found in the optical stimulus, since this assumption needs to be evaluated with respect to his whole theory of perception (Gibson,1966). It has become clear, however, that organizational and inferential processes affect perception strongly only when stimulus support is available.

Fluorescence

Evans (1959) has investigated the stimulus conditions that produce the perception of fluorescence. Observers viewed Munsell samples in front of a large white, gray, or black background. The perception of fluorescence was found to be a function of the gray content of the color. Evans found that observers could match Munsell samples for the amount of gray they contained. The amount of gray perceived in different Munsell samples was a function of the background. Less gray was perceived for gray and black backgrounds than for a white background. The gray component of a color also decreased with increasing luminance and with increasing purity. Evans reported that if the luminance of a Munsell sample of fixed purity is increased above the point of zero gray content (the color no longer appears to possess gray content), the sample takes on a fluorescent appearance. Further increase in the luminance of the sample produces the impression of a color in the illuminant mode. The perception of color in the illuminant mode occurred when the luminance of the sample was somewhat higher than the luminance of the background. Evans suggested that the illuminant mode of appearance occurs when the general adaptation level of the eye becomes controlled by the sample and not the background. The surface mode of appearance occurs when the luminance of the sample is below the general adaptation level established by the background. Evans proposed that fluorescence is an attribute of the surface mode. The general adaptation level is controlled by the background. He hypothesized that the perception of fluorescence occurs when a color receptor, or color receptors, are stimulated above the level that they are stimulated by the background. Phenomenally, fluorescence appears to be the name given to the perception of hues of high brightness and saturation in the surface mode.

Constancy in Photographs

Katz (*1911*, 1935) suggested that lightness constancy is not shown in photographs because a photograph depicts a scene on a flat physical surface. Though there are cues to the distribution of light in the scene, the perception of a paper's surface in a photograph works against seeing the colors in the picture in the illumination of the depicted scene. Pirenne (1970) has recently studied how the subsidiary awareness of a picture as a flat surface influences the perception of shape and depth in photographs and drawings. A comparable study of the perceptions of light and shade and of surface qualities such as glossiness and metallic colors has yet to be undertaken. There is no question that the tendency to constancy is weakened and that some constancy is lost in pictures. Because of the contradictory stimulation indicating that the image is flat, the interpretation of the variations in luminance as shadows and highlights on the child's face in Plate 6 is imperfect and a compromise results. Similarly, in Plate 7 an observer can take an attitude in which he sees the highlights as white spots on the picture's surface and in consequence loses the perception of glossiness.

Evans (1944) has pointed out that photography is a poor medium for representing differences in illumination. Shadows and highlights in pictures are, therefore, ordinarily seen as differences in lightness. Plate 8, in which the areas of reduced illumination are seen as gray, shows how completely lightness constancy may fail in a picture. However, Evans (1944) has shown that some degree of constancy is possible in photographs. The amount of constancy in a photograph depends on the extent to which the conditions of illumination and spatial arrangement of surfaces are convincingly presented. For example, Evans found almost no constancy of the background in a photograph because the distance of the background from the main figure in the photograph was not indicated. To compensate for the imperfect operation of the

Plate 8. A black-lighted plant. The areas of reduced illumination in this photograph are usually seen as areas of differing shades of gray rather than as parts of a plant that are in shadow. (Photograph courtesy of Eastman Kodak.)

constancy mechanisms, the photographer must arrange his lights or choose his angle so that contrast in "the lighting of the scene as viewed by the eye is very much less in all respects than is desired in the final picture" (Evans, 1944, p. 539). The photograph in Plate 9 represents a fairly satis-

Plate 9. A successful lighting of a rose. The areas of reduced illumination may be seen as white in shadow. (From R. M. Evans, *Eye, Film, and Camera in Color Photography* [New York: Wiley, 1959], p. 165; by permission of John Wiley & Sons, Inc. Photograph courtesy of Eastman Kodak.)

factory solution. The areas of reduced illumination may be seen as white in shadow.

Developmental Studies

Constancy does not depend on a highly developed intelligence. Chimpanzees, chickens, fish, and monkeys as well as humans have been shown to exhibit constancy (Locke, 1935). There is also apparently a high degree of constancy in young children. However, age differences may be expected to affect constancy in color perception when the perception depends upon identification, as it does in memory color, or upon the discrimination of minimal stimuli indicating light and shade. Age differences may also be expected to affect attitude: older observers, more often than children, adopt different attitudes; for example, they may change from an objective to an analytic attitude. Burzlaff (1931) used the method of illumination perspective to compare the perception of lightness by children and adults. Illumination perspective is the term used by Katz (*1911*, 1935) to describe the gradation of illumination due to the varying distances from a light source. Burzlaff placed one gray stimulus near a window and another gray stimulus in the back of the room so that its illumination was one-twentieth of the illumination falling on the gray near the window. Burzlaff found that lightness constancy increased from ages four to seven when single gray papers were presented in strongly and weakly illuminated areas of the room. He found, however, that the age trend disappeared when charts of gray papers ranging from white to black were used instead of single gray surfaces. Similar results were obtained by Brunswik (1929), also employing the method of illumination perspective to alter the luminance of single isolated surfaces. Beck (1966), who employed an experimental arrangement involving a cast shadow, likewise found a difference between the lightness constancy exhibited by adults and children. Both adults and children tended to exhibit constancy when a white surface was used, failed to exhibit constancy when the surface was black, and differed with gray surfaces, the adults exhibiting greater constancy than children.

IX

Conclusion

Our review has shown that there is as yet no general agreement on how an observer is able to perceive a stable color with changes in the intensity and spectral composition of the illumination. Scattered throughout the literature are diverse and conflicting interpretations of experiments on adaptation, contrast, and constancy. A review of a subject always presents a challenge to examine how the experimental findings can be integrated to provide an explanation of the facts presented. In this final chapter, I want to formulate my interpretation of the problems posed by the phenomenon of color constancy.

A fundamental theoretical question is whether or not the phenomena of color constancy and simultaneous contrast can be subsumed within a single explanatory principle. Aside from the influence of memory color, Helmholtz (*1867*, 1925) considered both color constancy and simultaneous contrast to be the result of a central transformation in which there is a change in the correspondence between a perceived color and the quality and intensity of a light stimulus; Hering considered them both to be the results of opponent physiological interactions (see Chapter III). More recently, Helson (1943) and Wallach (1948) attempted to subsume color

rather than favor constancy. Simultaneous contrast would be expected to oppose constancy when achromatic surfaces are surrounded by saturated chromatic surfaces. The effects of adaptation in successive contrast would be expected to oppose constancy when the eyes chromatically adapt in scanning a visual field. The experiment by Helson described in Chapter IV in which achromatic surfaces of high reflectance take on the color of the illuminant and achromatic surfaces of low reflectance take on the color complementary to that of the illuminant is an example of an extreme failure of constancy. Under ordinary circumstances, however, simultaneous contrast and successive contrast do not greatly disrupt constancy. Constancy is favored by the fact that a surface has a distinct texture, is spatially included within defined contours, and generally differs in depth from its immediate background. In Chapter III, these variables were shown to limit the effects of simultaneous and successive contrast and thereby lead to stable color perceptions. More attention needs to be paid to the fact that, if their effects were not inhibited, the processes of simultaneous contrast and adaptation which compensate for changes in the illumination and support color constancy would oppose color constancy with constant illumination when movement of a surface changes the surrounding colors or movement of the eyes project an afterimage on a surface. It is important to consider how this occurs. Logically, it is possible that there is a cortical inhibition of retinal simultaneous and successive contrast effects which would disrupt color constancy as a surface is moved in the visual field or the eye scans surfaces in the visual field. This cortical inhibition may depend on factors such as contours, texture, and figure-ground segregation which contribute to the object character of a percept. In Chapter III it was shown that there are processes which equalize contrast effects within a contoured area and within a figure-ground segregation. Experiments also indicated that color contrast decreases when a surface is moved out of the plane of its immediate background. A particular theory which distinguishes between the effects of a cortical adaptation and a retinal adaptation on constancy is that of Judd, described in Chapter V.

The second way in which sensory processes facilitate color constancy is that they transform a light stimulus into elements which provide a basis for the operation of perceptual mechanisms. For example, a sharp contour produces the neural correlates for a hard, well-defined surface. A fuzzy contour produces neural correlates for an ill-defined membrane or film appearance. The phenomenal terms employed to describe the result of the operations of sensory processes only characterize by analogy the information contained in the neural pattern prior to the operation of perceptual mechanisms. A membrane or film appearance favors interpreting the sensory signals in terms of a change in surface illumination rather than in terms of a change in surface color. Moreover, a change from a surface to a membrane appearance facilitates the perception of a duo-organization in which the surface is seen to extend behind the membrane. Both factors would tend to produce the perception of a shadow that lies on a surface. More generally, the operation of organizational processes depends upon unit formation produced by sensory processes. Sensory processes produce the neural correlates for the patterns of lightness and hue variations that are integrated in the perception of a glossy color, a metallic color, or the perception of a particular surface material such as wood, paper, or cloth.

What is the nature of the perceptual mechanisms? The results of experiments described in Chapters VI and VII show that the perceptions of illumination and lightness are not coupled psychologically in a one-to-one relation as implied by Helmholtz (*1867*, 1925) and Koffka (1935). The hypothesis of a precise psychological coupling was applied by Koffka also to the phenomenon of transparent colors. Koffka (1935) hypothesized that the perception of one colored surface behind another follows the laws of additive color mixture. When organizational processes lead to a duo-organization, an observer sees a yellow through a blue though these two component colors would ordinarily cancel each other and produce the appearance of a gray. The review of experimental results in Chapter VII, however, indicates that the phenomenon of transparent colors is labile and probably depends on specific experiences of how a color looks in an altered illumination. The suggestion that the phenomenon of transparency can occur when a surface is seen through a colored glass and a subtractive color mixture occurs would also indicate that it is not based on the splitting of a cortical color signal into its additive color mixture components. For example, light that reaches the eye from a bright saturated blue surface seen through a yellow glass is predominantly green. I find, however, that the observer can see the superposed area as a blue through a yellow when he takes an attitude which causes him to see the superposed area in two distinct planes. In the pair of circles at the bottom of the frontispiece, the perception of a blue through a transparent yellow color or the reverse is in accordance with the laws of subtractive color mixture.

I have suggested that the development of schemata for the perception of visual surfaces plays an important role in the perception of a surface color. The severance of a color of a surface from the color of the illumination depends upon the assimilation of sensory signals to an appropriate surface schema. The concept of a schema has been stressed as mediating the effects of past experience. It needs to be emphasized that the learning of a schema is strongly influenced by the processes which organize and integrate sensory signals. Organizational and integrative activity need to be postulated to explain the perception of surface textures and qualities described in Chapter VIII. The perception of overlapping, for example, that may occur with a cast shadow involves a perceptual tendency to resist the interruption and maintain the continuity of a surface. Processes of closure or completion also appear to be involved in the perception of a clear color such as that perceived in a pane of glass and in the perception of a glossy color. A continuous pane of glass is seen though there is stimulation occuring only at discrete points. Similarly, an entire surface is seen as glossy though highlights occur only at isolated points. What is suggested is that the brain first integrates and organizes sensory signals into more complex funtional units; these units are then encoded in terms of schemata, i.e., in terms of their central tendency and the ways in which they deviate from this central tendency. Sensory integrations of common patterns of stimulation become assimilated to a particular schema established through past experience. This explains the observation of Mach (*1886*, 1959) that "the visual sense acts in conformity with a principle of economy and, at the same time in conformity with a principle of probability" (p. 215).

Two factors influence the assimilation of sensory signals to schemata. First, there are organizational processes; second, there are cues to an altered illumination. These two factors have not always been clearly distinguished. The assimilation of sensory signals to schemata by both factors is strongly influenced by the attitude of the observer. An attitude in which an observer attends to the object and perceptual situation leads to greater constancy than an analytic or critical attitude in which an observer focuses his attention on the color quality of a surface area as distinct from the object and the perceptual situation (see Chapter IV).

In the operation of organizational processes, the perception of illumination is not a necessary condition for constancy but a parallel phenomenon accompanying constancy that arises from the underlying organizational processes. For example, the perception of a constant surface color and a varying illumination with an object shadow is a result of an organizational process that operates to minimize surface color differences. It is not that cues to the illumination lead to constancy, but rather that the tendency to perceive a uniform color leads to the perception of a difference in illumination. Another organizational principle suggested by the experiments of Kardos (see Chapter VII) is that illumination differences of surfaces in approximately the same plane are minimized. This principle derives from the tendency to apprehend the illumination in a scene as coming from a single light source that illuminates surfaces in the same plane equally. Organizational processes may also be involved in the operation of certain higher order variables. Regular patterns of change in luminance and chromaticity often are perceived as changes in illumination. In the experiment of Judd (see Chapter V), a stimulus consisting of a constant chromatic component added to a varying achromatic component gave rise to the perception of a difference in illumination. Gradations of illumination due to a varying distance from a light source also produce a gradient of luminance that is interpreted as an illumination change. More study is necessary of the patterns of luminance and chromaticity occurring in a scene that result in ascribing the color change to the illumination rather than to an object. Color constancy is greatest when the impression is that the illumination of the entire visual field has changed. For example, color constancy is strong if a colored glass is brought close to the eye so that the entire visual field is seen through the glass. There should be further investigation of the degree to which organizational factors may be involved in the operation of the higher order variables and of the variable of field size. The investigation should study whether or not color constancy tends to increase when the number of colors that may be seen as constant is increased by interpreting luminance and chromaticity changes as due to changes in the illumination.

The specific cues which give rise to the impression of an enhanced or lowered illumination are a second factor. What is important is the stimulus information that effectively creates the impression of a special illumination. The assimilation of sensory signals in terms of a schema appears to follow a principle of economy. There is a tendency to encode sensory signals in terms of a surface in the prevailing illumination. Only when there are specific cues that an achromatic surface is in a decreased or in an enhanced illumination is the surface seen as a white in shadow or as a black in spotlight rather than as a gray in the prevailing illumination. In general, the perceptual processes do not disregard relevant information. An observer's attention, however, may focus on stimulus features that are consistent with a particular schema. There are many stimulus variables to indicate that a surface is in a special illumination. Among these, the variable most studied is the properties of contours which indicate that an area of a surface is in reduced illumination (a cast shadow) or a spatial area is in reduced illumination (an air shadow). Though more study is required, Katz's (*1911*, 1935) and Koffka's (1935) hypothesis that luminance ratios greater than 60 to 1 give rise to an impression of a change in illumination rather than a change in surface color, if at all effective, appears to be a weak cue. Specific cues may operate in conjunction with organizational processes or independently of them. The principles of economy exhibited by the perceptual process may include that of reducing the discrepancy between a stimulus and a schematic representation that occurs frequently. The tendency to perceive a surface as being of a single color could have its basis in the fact that surfaces of uniform color occur frequently in our environment.

The defects of past experience may be mediated not only by cues to the illumination but also by memory images. Specific experiences of how a surface would look in an altered illumination may be important. I have suggested that the perception of transparency is facilitated by memory images of how a surface looks through a colored film. The perceptual mechanisms that determine the severance of illumination color from surface color, however, occur at an earlier level than that of recognition and identification. The experiences that are important in assimilating sensory signals to a schema are based not on the identification of a stimulus, but on specific stimulus features that are visually given. What is important are contour characteristics, highlights, and surface arrangements, and not cognitive information as such. Thus, color perception appears to be only secondarily affected by the identification of a stimulus that is involved in memory color experiments (see Chapter VII).

The discussion in Chapter VI emphasized that a schema does not establish a one-to-one relation between the perception of surface illumination and the perception of surface lightness. When a surface is uniformly illuminated and perceived to be uniformly illuminated, as in a dark-room setting, luminance variations are perceived as differences in surface lightness. Stimuli complex enough to give rise to the perception of a surface color, however, contain correlates for both the perceptions of the lightness and the illumination of a surface. The relation between the perceptions of lightness and illumination is a function of the relation between their stimulus correlates. The perceptions of illumination and lightness may or may not covary in the manner hypothesized by the albedo hypothesis, depending on their stimulus correlates. If perceived surface illumination is determined by the maximum intensity reflected from a visual field and perceived surface lightness is determined by the ratio of the luminance of a surface to the maximum luminance reflected from the visual field, then perceived lightness and illumination will covary in the manner predicted by the albedo hypothesis (see Chapter VI). However, if the stimulus correlate for the perceived surface illumination is the maximum luminance, but other luminances in the visual field are significant factors in determining perceived surface lightness, the perceived lightness and illumination will not covary in the manner predicted by the albedo hypothesis. When a surface is perceived to be illuminated nonuniformly, the perceptions of surface lightness, surface illumination, and the spatial position of a surface become interdependent. In Chapter VII we saw that decisions about whether an area of lower luminance is seen as a shadow or a difference in surface lightness may be tied to decisions about surface orientation and illumination direction. A variation in luminance may be seen as an area of reduced illumination or a gray spot, depending on the apparent orientation of the surface to the perceived direction of the illumination. Though a change in the apparent spatial position of a surface that alters the apparent illumination of a surface may alter perceived lightness, there is no necessary proportional variation between the altered perception of illumination and the altered perception of lightness. For example, stimulus cues indicating that a surface is strongly illuminated may preclude the perception of a white surface in a shadow and the surface will then be seen as gray. Sufficient stimulus support, however, may not be present for seeing the surface as a black under a heightened illumination and there is no proportional variation between the perceived darkening of a surface and the perception of a heightened illumination (see Chapter VII).

A surface which is made inconsistent with the overall direction of illumination may take on an unusual appearance. For example, the illuminated side of a Mach card after inversion may appear luminous; the shaded side may appear less substantial, softer, and the lightness variations become striking. I believe that these effects result when luminance gradients present on both the illuminated and shaded sides of the card are ascribed to the surface and not to the illumination. The luminance variations when seen as belonging to the surface change the appearance of the surface. The change that occurs in a percept when stimulus variations can not be ascribed to variations in the illumination can be seen in the photographs in Plate 6. Clarity of the lightness variations, I believe, is greater when the photograph is inverted because then the various gray areas can not be interpreted as lighted and shadowed areas due to the illumination and discounted as such.

The question of which properties of color are dimensions and vary continuously and which are discrete has not been completely resolved. Both continuous and discrete properties appear to exist. In Chapter II, I suggested that modes of color appearance are discrete. There are no true intermediaries between the film, surface, and illuminant modes. A conceptualization in terms of continuous dimensions would seem to be appropriate for color attributes that do not depent on integrative and organizational processes. For example, the perceptions of hue, saturation, brightness, fluorescence, and transparency do not depend on the integration of separate elements, and these perceptions vary continuously. But when the perception of attributes depends upon the integration of differences in luminance and chromaticity from various parts of a surface, as does the perception of glossy and metallic colors, the question is more difficult. The glossy to matte dimension is continuous. The degree of perceived gloss depends on the degree to which a surface departs from a mirror surface. Whether the difference between metallic and nonmetallic colors, for example between a silver color and a gray color, allows for a continuous transition is a debatable question at present. In Chapter VI, a definition of the pronouncedness of achromatic surface colors was proposed in terms of alternative interpretations of sensory signals. As the illumination of a white surface is decreased, the surface takes on a dark cast; the cast, however, is ascribed to the illumination, and the surface is seen as a white surface of decreased pronouncedness. As the illumination on a black surface is increased, the surface takes on a light cast; the cast, however, is ascribed to the illumination, and the surface is seen as a black surface of decreased pronouncedness. According to this definition, the attribute of pronouncedness allows for continuous variation and constitutes a dimension.

The analysis presented is that perceptual mechanisms involve both inferential and organizational processes. Though I have used the word "inference," this is not the best term. There is no conscious formulation of rules, examination of premises, or checking of inferences; there are no conscious intermediate stages. The term "assimilation" has been used metaphorically to suggest the kind of processing involved.

The matching of sensory signals to a schema is rapid, automatic, and resists modification when reinforced a number of times by experience. Another term is really needed to distinguish this type of inference from the explicit and deliberate type of inference which occurs in thought. The stereotyped nature of perceptual processing and its dependence on specific stimulus cues indicates that the processing depends on the development of specific mediating mechanisms such as schemata and not on a general cognitive inference.

A final problem that we must briefly turn to is that of conscious awareness. In Chapter VI it was suggested that information about the absolute luminance of a stimulus is not available to awareness when lateral inhibitory interactions alter the neural signal. Though this information is not available to awareness, it may still affect some perceptual judgments. The hypothesis that has been proposed is that the perception of color is changed by the assimilation of sensory signals to different schemata. However, an observer with an analytic attitude may become aware of a color property that remains constant in spite of his assimilation of sensory signals to different schemata. An intensive quality, for example, appears to remain constant in the transformation from a white of reduced illumination or pronouncedness to a gray as a result of a change in an observer's attitude (see Chapter VI). To what extent this is true with the transformation in appearance resulting from penumbral cues, or changes in the apparent position of a surface as in object shadows and transparent colors is not clear. There are probably differing levels of awareness and individual differences in this matter. When an observer is asked, "What do you really see?" he appears to become aware of the sensory pattern underlying complex color attributes. He no longer apprehends as an integrated whole the pattern of lightness and hue variations that produces the perception of a glossy color or of a metallic color; he now sees the separate elements constituting the pattern. The mental attitude adopted when such a question is posed interferes with the operation of integrative processes and increases awareness of the sensory signals. This is especially true in looking at pictures because of the contradictory stimulation arising from the physically flat surfaces (see Chapter VIII).

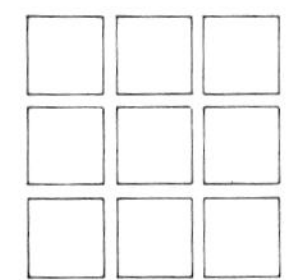

Glossary

Achromatic color. Color not possessing hue; self-luminous achromatic colors range from dim to bright, and surface achromatic colors range from black to white.

Achromatic point. The chromaticity of a color expressed in terms of trichromatic coefficients (a point in the chromaticity diagram) that is seen as achromatic.

Adaptation level. A weighted geometric mean of the luminances in the field of view; reflectances above the adaptation level tend toward the hue of the illuminant, and those below the adaptation level toward the hue of the complementary afterimage.

Afterimage. An aftereffect following the observation of a color stimulus; a positive afterimage is the same color as the stimulus; a negative afterimage is the complementary color of the stimulus.

Air shadow. A tridimensional shadow that is seen to fill a spatial region; an example is a shadowed corner.

Air-shadow setting. An experimental paradigm for studying color constancy in which a target in good illumination is compared to a target in a spatial region of reduced illumination due to an air shadow.

Albedo hypothesis. The view that the intensity of the illumination is discounted in perceiving lightness (see p. 99).

Apparent luminance. The intensive magnitude of a surface color that is presumed to remain constant with changes in lightness resulting from perceptual mechanisms.

Assimilation. A color effect opposite to that which occurs in simultaneous contrast; light areas lighten a gray background and dark areas darken a gray background.

Attributes. The qualities that may be predicated of colors in a given mode of appearance.

Bipolar cells. A layer of retinal cells which receive stimulation from the rods and cones and transmit impulses to the ganglion cells.

Brightness. The attribute that varies for self-luminous colors from very dim to very bright; in connection with surface colors, see pp. 8–15.

Brunswik ratio. A measure of constancy (see p. 58).

Candle. A unit of luminous intensity; one-sixtieth of the luminous intensity of one square centimeter of a black body radiator operated at the temperature of freezing platinum.

Cast shadow. A darkened area on a surface from which light rays have been blocked by interposing an object between the surface and the light source; a cast shadow may be seen as lying on a surface and is distinguished by its fuzzy edge or penumbra.

Chromatic adaptation. Adjustment of the sensitivity of the eye to the chromaticity of the illumination; a decrease in sensitivity of the mechanisms most strongly stimulated by the chromatic illumination and an increase in sensitivity of the remaining mechanisms.

Chromatic colors. Colors possessing hue.

Chromaticity. The aspect of a color determined by its dominant wave length and purity taken together.

Chromaticity diagram. A representation of chromaticities in a plane figure obtained by plotting colors in terms of two of the three trichromatic coefficients.

Closure. *See* Gestalt laws.

Color. The property of light by which two objects of the same size, shape, and texture can be distinguished; includes both chromatic and achromatic color.

Color constancy. The approximate constancy in the perceived color of an object despite changing illumination that alters the intensity and spectral composition of the light stimulating the eyes; lightness constancy refers to the constancy of lightness with changes in the intensity of the illumination, and hue constancy refers to the constancy of hue with changes in the spectral composition of the illumination.

Color transformation. The discounting of the illumination color due to a central adjustment of the chromatic neutral point (i.e., the stimulus seen as achromatic); to be distinguished from the effects due to simultaneous and successive contrast.

Colorimetric purity. The ratio of the luminance of a spectrum color to the luminances of the spectrum color and an achromatic color that when mixed match a given color.

Complementary colors. Two colors that when added together in suitable proportions yield an achromatic color.

Dark adaptation. *See* Light-dark adaptation.

Dark-room setting. An experimental paradigm for studying color constancy in which the target and background surfaces are perceived in an otherwise totally dark room; they are uniformly illuminated and are perceived to be uniformly illuminated.

Diffuse reflection. Reflection of light from a rough or matte surface in which light rays are reflected in many directions for any given angle of incidence.

Dimension. An attribute that varies continuously between two values; the dimension of hue varies in a circle from red through yellow, green, blue, violet, and purple to red again; the dimension of brightness varies along a straight line from just visible to dazzling.

Dominant wavelength. Wavelength of a spectrum color which, when added to an achromatic color, will match a given color.

Duo-organization. The perception of one surface behind another surface, as in transparent colors.

Excitation purity. The ratio of the distance between the achromatic point in the chromaticity diagram and a point representing a color to the distance between the achromatic point and the dominant wavelength of the color.

Figure-ground perception. The perception of a portion of a visual pattern as foreground and the remainder as the background; foregrounding may occur on several levels, as A is foregrounded with respect to B, B with respect to C, and the like.

Film color. Color perceived as an expanse of light in a bidimensional plane without objective reference.

Foot-candle (ft.-c.). A unit of illuminance; the illuminance on a surface every point of which is one foot away from a one-candle point source.

Foot-Lambert (ft.-L.). A unit of luminance; a perfectly reflecting diffuse surface has a luminance of one foot-Lambert when illuminated by one foot-candle.

Ganglion cells. A layer of retinal cells which receive stimulation from the bipolar cells and transmit impulses to the optic nerve.

Ganzfeld. Conditions which produce complete homogeneous retinal stimulation.

Gestalt laws. Rules which describe how a stimulus object is organized; the *Law of Prägnanz* states that perceptual mechanisms lead to a representation that is "best" (minimizes complexity) of the many representations that are consistent with the pattern of stimulation; closure is the tendency to perceive unified or whole figures; grouping by similarity and proximity is the tendency to organize a visual pattern into sub units in accordance with these factors.

Glossiness. An attribute of matte to glossy surfaces; various kinds of glossiness correlate with different patterns of specular and diffuse reflection (see Table 2).

Grouping. *See* Gestalt laws.

Higher order variables. Variables of the adjacent and successive order of wavelengths and intensities of a stimulus (gradients) that are in psychophysical correspondence with phenomenal properties.

Hue. The attribute that allows colors to be classified as red, yellow, green, blue, or intermediates of these.

Hue constancy. *See* Color constancy.

Illuminant color. Color perceived as a source of light.

Illumination color. Color perceived as a property of illumination rather than as a property of an object or of a film.

Illumination frame of reference. An internal representation of the intensity and spectral composition of the illumination in a spatial region which operates to determine the colors of objects in the spatial region.

Inferential processes. Processes which derive the perceptual characteristics of a stimulus from cues; the search and interpretation of cues is normally unconscious.

Insistence. Defined by Katz as the degree to which a color forces itself upon attention.

Lateral inhibition. A decrease in the neural intensity signal (frequency of impluses) from a receptor when neighboring receptors are stimulated.

Laws of Field Size. Katz (*1911*, 1935) hypothesized that increasing the size of an area of altered luminance or chromaticity tends to produce the impression of a change in illumination rather than of a change in surface color; first law of Field Size refers to a change in retinal size; second law of Field Size refers to a change in apparent size (retinal size held constant).

Law of Prägnanz. See Gestalt laws.

Light. *See* Visible spectrum.

Light-dark adaptation. Adjustment of the sensitivity of the eye to the prevailing intensity of the illumination; exposure to light (light adaptation) decreases sensitivity; exposure to dark (dark adaptation) increases sensitivity.

Lightness. The attribute which varies for surface and volume colors from black through gray to white.

Lightness constancy. *See* Color constancy.

Luminance. A measure of the effectiveness of radiant energy in stimulating the eye; the luminance of a sample of light is the radiant energy of the sample weighted by the photopic luminosity function.

Mach bands. Contrast effect, reflecting local excitation and inhibition interactions that enhance or sharpen a blurred or fuzzy contour. An abrupt increase in light intensity produces a dark band, and an abrupt decrease in light intensity produces a light band.

Maxwell triangle. A spatial representation of colors in terms of three primaries (reference colors) located at the vertices of an equilateral triangle, the position of a color in the triangle indicates the relative amounts of the three primaries that when mixed match the color.

Memory color. The color of a familiar object tends to be in accordance with its daylight color.

Metallic colors. Characteristic colors associated with the particular appearance of metals; gold, silver, copper and the like.

Millimicron ($m\mu$). A unit of distance used to describe wavelength; one millimicron is equal to one millionth of a millimeter.

Mirror color. The reflection of a color in a polished surface in which the color of the object is seen behind and through the color of the reflecting surface.

Modes of appearance. The ways in which colors are experienced; color as a property of a film, a surface, illumination, and the like.

Munsell chroma. A scale indicating the saturation of a color ranging in approximately equal visual steps from /0 (a neutral gray of the same value) to /10 or farther (the maximum obtainable saturation for the given sample).

Munsell hue. A scale indicating the hue of a color ranging in approximately equal visual steps from red through purple.

Munsell value. A scale indicating the lightness of a color ranging in approximately equal visual steps from 0/ (ideal black) to 10/ (ideal white).

Negative afterimage. *See* Afterimage.

Neutral color. Same as achromantic color.

Nonselective samples. Samples which reflect light uniformly or approximately uniformly throughout the visible spectrum; nonselective samples appear achromatic.

Object shadow. The shadow lies on an object and is created by the shape and spatial orientation of the object to the light source; an example of object shadows are the numerous shadows present on a crumpled towel which are not seen as shadows but as folds and creases of the towel.

Organizational processes. Processes in which properties of the whole process determine the representation of a stimulus; for example, the organization of a stimulus to minimize color differences or shape differences.

Perceptual mechanisms. Representational mechanisms such as schemata and frames of reference that generate a description of a visual stimulus; perceptual mechanisms may involve both inferential and organizational processes.

Photopic luminosity function. The sensitivity of the eye under light adaptation to wavelengths between 400 and 700 mμ; shows the sensitivity of the cones to light.

Positive afterimage. *See* Afterimage.

Pronouncedness. Defined by Katz as the degree of goodness of a color—the whiteness of white or the redness of red.

Purity. The relative position of a color between an achromatic color and a spectral color of the same dominant wavelength;

two measures of purity are colorimetric purity and excitation purity.

Saturation. The attribute that determines the degree of difference of a color from a gray of the same lightness.

Schema. An internal representation of the central tendency or communality among a class of stimuli; the assimilation of sensory signals to a schema involves both inferential and organizational processes.

Scotopic luminosity function. The sensitivity of the eye under dark adaptation to wavelengths between 400 and 700 mμ; shows the sensitivity of the rods to light.

Selective samples. Samples which reflect certain wavelengths strongly, selective samples are seen as red, yellow, green, blue, and the like.

Sensory processes. Processes such as adaptation, contrast, and contour formation that determine a central neural pattern in accordance with the luminance distribution; these provide a basis for the utilization of perceptual mechanisms.

Sensory signals. The central neural pattern resulting from the operation of sensory processes.

Simultaneous contrast. Changes in either the lightness or hue of a target produced by its surround; if an achromatic target is surrounded by a background of higher luminance, the target is darkened, if by a background of a particular hue, the target takes on the complementary hue.

Spatial summation. Two light stimuli applied to neighboring retinal areas will summate in their effects and decrease the threshold for luminance.

Spectral composition. The distribution of energy as a function of wavelength between 400 and 700 mμ; also called wavelength composition.

Spectral energy distribution. *See* Spectral composition.

Specular reflection. Reflection of light from a smooth surface like a mirror in which the angle at which a light ray is reflected from the surface is equal to the angle at which the light ray strikes the surface, measured with respect to a line perpendicular to the surface.

Successive contrast. Changes in either the lightness or hue of a target resulting from the color just previously seen; a consequence of superimposing different stimuli on the same retinal region.

Surface color. Color perceived as a property of a surface.

Thouless ratio. A measure of constancy (see p. 58).

Transparent colors. Colors that can be simultaneously perceived, one behind the other.

Trichromatic coefficients. A way of specifying a color by stating the relative amounts of three primaries (reference colors) that when mixed match the color.

Uniform chromaticity triangle. Maxwell triangle in which the distances between points correspond approximately to the perceived differences in the chromaticities represented by the points.

Visible spectrum. The range of radiant energy between 400 and 700 millimicrons to which the eye is sensitive.

Volume color. Color perceived as a property of a volume; the colors are at least partially transparent and objects may be seen through them.

Wavelength. The distance from the crest of one wave to that of the next; the lengths of light waves are ordinarily expressed in the metric system (*see* Millimicron).

Wavelength composition. *See* Spectral composition.

Weber fraction. The ratio of the just noticeable difference of a stimulus to the total intensity at which the j.n.d. was obtained; the Weber fraction is constant in the middle range of intensities.

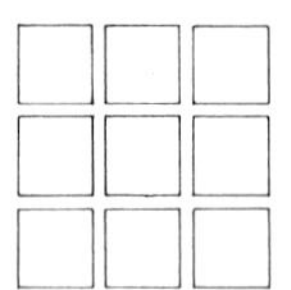

Bibliography

Avant, L. L. 1965. Vision in the Ganzfeld. *Psychol. Bull.,* 64: 246–258.

Bartelson, C. J., and Breneman, E. J. 1967. Brightness perception in complex fields. *J. Opt. Soc. Amer.,* 57: 953–957.

Bartlett, F. C. 1932. *Remembering.* Cambridge: Cambridge University Press.

Beck, J. 1959. Stimulus correlates for the judged illumination of a surface. *J. Exp. Psychol.,* 58: 267–274.

——. 1961. Judgments of surface illumination and lightness. *J. Exp. Psychol.,* 61: 368–373.

——. 1964. The effect of gloss on perceived lightness. *Amer. J. Psychol.,* 77: 54–63.

——. 1965. Apparent spatial position and the perception of lightness. *J. Exp. Psychol.,* 69: 170–179.

——. 1966a. Contrast and assimilation in lightness judgments. *Percep. and Psychophy.,* 1: 342–344.

——. 1966b. Age differences in lightness perception. *Psycho. Sci.,* 4: 201–202 (comment, 5: 166).

——. 1969. Lightness and orientation. *Amer. J. Psychol.,* 82: 359–366.

——. 1971. Surface lightness and cues for the illumination. *Amer. J. Psychol.,* 84: 1–11.

——. Forthcoming. Dimensions of an achromatic surface color. In *Studies in perceptual theory,* ed. H. L. Pick, R. B. MacLeod, and J. Kennedy. Ithaca: Cornell University Press.

Békésy, G. von. 1968a. Brightness distribution across the Mach bands measured with flicker photometry and the linearity of sensory nervous interaction. *J. Opt. Soc. Amer.*, 58: 1–8.

——. 1968b. Mach- and Hering-type lateral inhibition in vision. *Vision Res.*, 8: 1483–1499.

Belsey, R. 1964. Color perception and the Land two-color projection. *J. Opt. Soc. Amer.*, 54: 529–531.

Benary, W. 1924. Beobachtungen zu einem Experiment über Helligkeitskontrast. *Psychol. Forsch.*, 5: 131–142.

Berman, P., and Leibowitz, H. W. 1965. Some effects of contour on simultaneous brightness contrast. *J. Exp. Psychol.*, 69: 251–256.

Bixby, F. L. 1928. A phenomenological study of luster. *J. Gen. Psychol.*, 1: 136–174.

Bolles, R. C., Hulicka, I. M., and Hanly, B. 1959. Color judgments as a function of stimulus conditions and memory color. *Canad. J. Psychol.*, 13: 175–185.

Boring, E. G. 1942. *Sensation and perception in the history of experimental psychology*. New York: Appleton Century Crofts.

Brown, J. L., and Mueller, C. B. 1965. Brightness discrimination and brightness contrast. In *Vision and visual perception*, ed. C. H. Graham, pp. 208–250. New York: Wiley.

Bruner, J. S., and Postman, L. 1949. On the perception of incongruity: a paradigm. *J. Pers.*, 18: 206–223.

——, and Rodrigues, J. 1951. Expectation and the perception of color. *Amer. J. Psychol.*, 64: 216–227.

Brunswik, E. 1928. Zur Entwicklung der Albedowahrnehmung. *Z. Psychol.*, 109: 40–115.

——. 1956. *Perception and the representative design of psychological experiments*. Berkeley: Univ. of Calif. Press.

Burgh, P., and Grindley, G. C. 1962. Size of test patch and simultaneous contrast. *Quart. J. Exp. Psychol.*, 14: 89–93.

Burnham, R. W. 1953. Bezold's color mixture-effect. *Amer. J. Psychol.*, 66: 377–385.

Burzlaff, W. 1931. Methodologische Beiträge zum Problem der Farbenkonstanz. *Z. Psychol.*, 119: 117–235.

Committee on Colorimetry. 1953. *The science of color*. New York: Thomas Y. Crowell.

Coren, S. 1969. Brightness contrast as a function of figure-ground relations. *J. Exp. Psychol.*, 80: 517–524.

Cornsweet, T. N. 1970. *Visual perception*. New York: Academic Press.

Delk, J. L., and Fillenbaum, S. 1965. Differences in perceived color as a function of characteristic color. *Amer. J. Psychol.*, 78: 290–293.

De Valois, R. L., and Pease, P. L. 1971. Contours and contrast: responses of monkey lateral geniculate nucleus cells to luminance and color figures. *Science*, 171: 694–696.

Diamond, A. L. 1953. Foveal simultaneous brightness contrast as a function of inducing- and test-field luminances. *J. Exp. Psychol.*, 45: 304–314.

——. 1955. Foveal simultaneous contrast as a function of inducing field area. *J. Exp. Psychol.*, 50: 144–152.

——. 1962. Simultaneous contrast as a function of test-field area. *J. Exp. Psychol.*, 64: 336–345.

Dowling, J. E. 1967. The site of visual adaptation. *Science*, 155: 273–279.

Duncker, K. 1939. The influence of past experience upon perceptual properties. *Amer. J. Psychol.*, 52: 255–265.

Dunn, B., and Leibowitz, H. 1961. The effect of separation between test and inducing fields on brightness constancy. *J. Exp. Psychol.*, 61: 505–507.

Evans, R. M. 1944. Brightness constancy in photographic reproductions. *J. Opt. Soc. Amer.*, 34: 533–540.

——. 1948. *An introduction to color*. New York: Wiley.

——. 1959a. Fluorescent and gray content of surface colors. *J. Opt. Soc. Amer.*, 49: 1049–1059.

——. 1959b. *Eye, film, and camera in color photography*. New York: Wiley.

——. 1964. Variables of perceived color. *J. Opt. Soc. Amer.*, 54: 1467–1474.

——. 1967. Luminance and induced colors from adaptation to

100-millilambert monochromatic light. *J. Opt. Soc. Amer.*, 57: 279–281.

Festinger, L., Coren, S., and Rivers, G. 1970. The effect of attention on brightness contrast and assimilation. *Amer. J. Psychol.*, 83: 189–207.

Fisher, C., Hull, C., and Holtz, P. 1956. Past experience and perception: memory color. *Amer. J. Psychol.*, 69: 546–560.

Flock, H. R. 1970. Jameson and Hurvich's theory of brightness contrast. *Percep. and Psychophy.*, 8: 118–124.

——, and Freedberg, E. 1970. Perceived angle of incidence and achromatic surface color. *Percep. and Psychophy.*, 8: 251–256.

Freeman, R. B. 1967. Contrast interpretations of brightness constancy. *Psychol. Bull.*, 67: 165–187.

Fry, G. A. 1931. The stimulus correlates of bulky color. *Amer. J. Psychol.*, 43: 618–620.

Fry, G. A., and Alpern, M. 1953. The effect of a peripheral glare source upon the apparent brightness of an object. *J. Opt. Soc. Amer.*, 43: 189–195.

Fuchs, W. 1923. Experimentelle Untersuchungen über das simultane Hintereinandersehen auf der selben Sehrichtung. *Z. Psychol.*, 91: 145–235.

Gelb, A. 1929. Die "Farbenkonstanz" der Sehdinge. In *Handbuch der normalen und pathologischen Physiologie,* ed. A. Bethe, 12: 594–678. Berlin: Springer.

Gibson, J. J. 1950a. The perception of visual surfaces. *Amer. J. Psychol.*, 63: 367–384.

——. 1950b. *The perception of the visual world*. Boston: Houghton Mifflin.

——. 1966. *The senses considered as perceptual systems*. Boston: Houghton Mifflin.

——, Purdy, J., and Lawrence, L. 1955. A method of controlling stimulation for the study of space perception: the optical tunnel. *J. Exp. Psychol.*, 50: 1–14.

Gogel, W. C., and Mershon, D. H. 1969. Depth adjacency in simultaneous contrast. *Percep. and Psychophy.*, 5: 13–17.

Haber, R. N. 1965. Effect of prior knowledge of the stimulus on word recognition process. *J. Exp. Psychol.,* 69: 282–286.

Hanawalt, H. G., and Post, B. E. 1942. Memory trace for color. *J. Exp. Psychol.,* 30: 216–227.

Harper, R. S. 1953. The perceptual modification of colored figures. *Amer. J. Psychol.,* 66: 86–89.

Heider, G. M. 1932. New studies in transparency, form and color. *Psychol. Forsch.,* 17: 13–55.

Heinemann, E. G. 1955. Simultaneous brightness induction as a function of inducing- and test-field luminances. *J. Exp. Psychol.,* 50: 89–96.

——. 1961. The relation of apparent brightness to the threshold for differences in luminance. *J. Exp. Psychol.,* 61: 389–399.

Helmholtz, H. von. *1867,* 1924–1925. *Helmholtz's treatise on physiological optics.* Edited by J. P. Southall, translated from the 3d German edition. New York: Optical Society of America.

——. *1868,* 1962. The recent progress of the theory of vision. In *Popular scientific lectures,* ed. M. Kline, pp. 93–185. New York: Dover.

Helson, H. 1938. Fundamental problems in color vision I. The principle governing changes in hue, saturation, and lightness of non-selective samples in chromatic illumination. *J. Exp. Psychol.,* 23: 439–476.

——. 1943. Some factors and implications of color constancy. *J. Opt. Soc. Amer.,* 33: 555–567.

——. 1964. *Adaptation-level theory.* New York: Harper and Row.

——, and Jeffers, V. B. 1940. Fundamental problems in color vision II. Hue, brightness, and saturation of selective samples in chromatic illumination. *J. Exp. Psychol.,* 26: 1–27.

——, and Judd, D. B. 1932. A study in photopic adaptation. *J. Exp. Psychol.* 15: 380–398.

Henneman, R. H. 1935. A photometric study of the perception of object color. *Arch. Psychol.,* no. 179.

Hering E. *1874,* 1964. *Outlines of a theory of the light sense.* Translated from the German by L. M. Hurvich and D. Jameson. Cambridge, Mass.: Harvard University Press.

Hess, C., and Pretori, H. 1894. Messende Untersuchungen über die Gesetzmäsigkeit des simultanen Helligkeits-Contrastes. *Arch. Opthal.,* 40: 1–27. Translated by H. R. Flock, and J. H. Tenney. 1969. *Technical Report FLP-1. York University.*

Hochberg, J. E., and Beck, J. 1954. Apparent spatial arrangement and perceived brightness. *J. Exp. psychol.,* 47: 263–266.

——, Treibel, W., and Seaman, G. 1951. Color adaptation under conditions of homogeneous visual stimulation (Ganzfeld). *J. Exp. Psychol.,* 41: 153–159.

Horeman, H. W. 1963. Inductive brightness depression as influenced by configurational conditions. *Vision Res.,* 3: 121–130.

Hsia, Y. 1943. Whiteness constancy as a function of difference in illumination. *Arch. Psychol.,* no. 284.

Hunter, R. S. 1937. Method of determining gloss. *J. Research Natl. Bur. Std.,* 18: 19–38.

Hurvich, L., and Jameson, D. 1966. *The perception of brightness and darkness.* Boston: Allyn and Bacon.

James, W. 1890. *Principles of psychology.* New York: Holt.

Jameson, D., and Hurvich, L. M. 1961. Complexities of perceived brightness. *Science,* 133: 174–179.

——. 1964. Theory of brightness and color contrast in human vision. *Vision Res.,* 4: 135–154.

——. 1970. Improvable, yes; insoluble, no: a reply to Flock. *Percep. and Psychophy.,* 8: 125–128.

Jones, L. A., and Higgins, G. C. 1947. Photographic granularity and graininess: III. Some characteristics of the visual system of importance in the evaluation of graininess and granularity, *J. Opt. Soc. Amer.,* 37: 217–263.

Judd, D. B. 1940. Hue, saturation, and lightness of surface colors with chromatic illumination. *J. Opt. Soc. Amer.,* 30: 2–32.

——. 1941. The definition of black and white. *Amer. J. Psychol.,* 54: 289–294.

——. 1960. Appraisal of Land's work on two-primary color projections. *J. Opt. Soc. Amer.,* 50: 254–267.

——, and Wyszecki, G. 1963. *Color in business, science, and industry.* New York: Wiley.

Kanizsa, G. 1969. Perception, past experience and the impossible experiment. *Acta Psychologica,* 31: 66–96.

Kardos, L. 1928. Dingfarbenwahrnehmung und Duplizitätstheorie. *Z. Psychol.,* 108: 240–314.

——. 1934. Ding und Schatten. *Z. Psychol.,* Ergbd. no. 23.

Katona, G. 1929. Zur Analyse der Helligkeitskonstanz. *Psychol. Forsch.,* 12: 94–126.

——. 1935. Color contrast and color constancy. *J. Exp. Psychol.,* 18: 49–63.

Katz, D. *1911,* 1935. *The world of color.* Translated from the 2d German edition by R. B. MacLeod and C. W. Fox. London: Kegan Paul, Trench, Trubner.

Kinney, J. A. 1962. Factors affecting induced color. *Vision Res.,* 2: 503–525.

Kodak Color Data Books. 1950. *Color as seen and photographed.* Rochester: Eastman Kodak.

Koffka, K. 1932. Some remarks on the theory of color constancy. *Psychol. Forsch.,* 16: 329–353.

——. 1935. *Principles of gestalt psychology.* New York: Harcourt Brace.

——, and Harrower, M. R. 1931. Color and organization, I. *Psychol. Forsch.,* 15: 145–192.

Kozaki, A. 1963. A further study in the relationship between brightness constancy and contrast. *Jap. Psychol. Res.,* 5: 129–136.

——. 1965. The effect of co-existent stimuli other than the test stimulus on brightness constancy. *Jap. Psychol. Res.,* 7: 138–147.

Krauskopf, J. 1963. Effect of retinal image stabilization on the appearance of heterochromatic targets. *J. Opt. Soc. Amer.,* 53: 741–744.

Land, E. H. 1951. Color vision and the natural image, Parts I and II. *Proc. Nat. Acad. Sci.,* 45: 115–129, 636–644.

——. 1959. Experiments in color vision. *Scientific American,* 200 (5), 84–97.

——. 1964. The retinex. *American Scientist,* 52: 247–264.

Landauer, A. A. 1962. The D ratio: a new method for the measure of brightness constancy. *Australian J. Psychol.*, 14: 22–25.
——, and Rodger, R. S. 1964. Effect of "apparent" instructions on brightness judgments. *J. Exp. Psychol.*, 68: 80–84.
Lauenstein, L. 1938. Über Räumliche Wirkungen von Licht und Schatten. *Psychol. Forsch.*, 22: 267–319.
Leeper, R. W. 1935. A study of a neglected portion of the field of learning—the development of sensory organization. *J. Genet. Psychol.*, 46: 41–75.
Leibowitz, H. 1956. Relation between the Brunswik and Thouless ratios and functional relations in experimental investigations of perceived shape, size, and brightness. *Percept. Mot. Skills*, 6: 65–68.
——, and Chinetti, P. 1957. Effect of reduced exposure duration on brightness constancy. *J. Exp. Psychol.*, 54: 49–53.
——, Mote, F. A., and Thurlow, W. R. 1953. Simultaneous contrast as a function of separation between test and inducing fields. *J. Exp. Psychol.*, 46: 453–456.
——, Myers, N. A., and Chinetti, P. 1955. The role of simultaneous contrast in brightness constancy. *J. Exp. Psychol.*, 50: 15–18.
Lie, M. 1969a. Psychophysical invariants of achromatic colour vision I. The multidimensionality of achromatic colour experience. *Scandinavian J. Psychol.*, 10: 167–175.
——, 1969b. Psychophysical invariants of achromatic colour vision IV. Depth adjacency and simultaneous contrast. *Scandinavian J. Psychol.*, 10: 282–286.
Locke, N. M. 1935. Color constancy in the rhesus monkey and in man. *Arch. Psychol.*, no. 193.
MacAdam, D. L. 1961. A nonlinear hypothesis for achromatic adaptation. *Vision Res.*, 1: 9–41.
——. 1963. Chromatic adaptation. II. nonlinear hypothesis. *J. Opt. Soc. Amer.*, 53: 1441–1445.
Mach, E. *1886*, 1959. *The Analysis of Sensations*. Translated from the 5th German edition by S. Waterlow. New York: Dover.

MacLeod, R. B. 1932. An experimental investigation of brightness constancy. *Arch. Psychol.,* no. 135.

——. 1940. Brightness constancy in unrecognized shadows. *J. Exp. Psychol.,* 27: 1–22.

——. 1947. The effects of artificial penumbrae on the brightness of included areas. In *Miscellanea Psychologica Albert Michotte,* pp. 138–154. Louvain: Institut Superieur de Philosophie.

Marimont, R. B. 1962. Model for visual response to contrast. *J. Opt. Soc. Amer.,* 52: 800–806.

Martin, M. F. 1922. Film, surface and bulky colors and their intermediates. *Amer. J. Psychol.,* 33: 451–480.

Mershon, D. H., and Gogel, W. C. 1970. Effect of stereoscopic cues on perceived whiteness. *Amer. J. Psychol.,* 83: 55–67.

Metelli, F. 1970. An algebraic development of the theory of perceptual transparency. *Ergonomics,* 13: 59–66.

Michels, W. C., and Helson, H. 1949. A reformulation of the Fechner Law in terms of adaptation level applied to rating scale data. *Amer. J. Psychol.,* 62: 355–368.

Mikesell, W. H., and Bentley, M. 1930. Configuration and contrast. *J. Exp. Psychol.,* 13: 1–23.

Newhall, S. M. 1942. The reversal of simultaneous brightness contrast. *J. Exp. Psychol.,* 31: 393–409.

——, Burnham, R. W., and Clark, J. R. 1957. Comparison of successive with simultaneous color matching. *J. Opt. Soc. Amer.,* 47: 43–55.

——, ——, and Evans, R. M. 1958. Color constancy in shadows. *J. Opt. Soc. Amer.,* 48: 976–984.

Newson, L. J. 1958. Some principles governing changes in the apparent lightness of test surfaces isolated from their normal background. *Quart. J. Exp. Psychol.,* 10: 82–95.

O'Brien, V. 1958. Contour perception, illusion and reality. *J. Opt. Soc. Amer.,* 48: 112–119.

Oyama, T. 1968. Stimulus determinants of brightness constancy and the perception of illumination. *Jap. Psychol. Res.,* 10: 146–155.

Pearson, D. E., Rubenstein, C. B., and Spivack, G. J. 1969.

Comparison of perceived color in two-primary computer-generated artificial images with predictions based on the Helson-Judd formulation. *J. Opt. Soc. Amer.*, 59: 644–658.
Perky, C. W. 1910. An experimental study of imagination. *Amer. J. Psychol.*, 21: 422–452.
Petter, G. 1956. Nuove ricerche sperimentali sulla totalizzazione percettiva. *Riv. Psycho.*, 50: 213–227.
Pickett, R. M. 1968. Perceiving visual texture; a literature survey. *Aerospace Medical Research Laboratory Report, Wright-Patterson Air Force Base, no. AMRL-TR-68-12.*
Pirrene, M. H. 1970. *Optics, painting, and photography*. Cambridge: Cambridge University Press.
Prentice, W. C. H., Krimsky, J., and Barker, S. 1951. The roles of pattern and apparent distance in determining the colors of areas seen through transparencies. *J. Exp. Psychol.*, 42: 201–206.
Ratliff, F. 1962. Some interrelations among physics, physiology, and psychology in the study of vision. In *Psychology: a study of a science,* ed. S. Koch, Study II, 4: 417–482. New York: McGraw Hill.
Richards, W., and Parks, E. 1971. Model for color conversion. *J. Opt. Soc. Amer.*, 61: 971–976.
Rushton, W. A. H., and Westheimer, G. 1962. The effect upon the rod threshold of bleaching neighboring rods. *J. Physiol.*, 164: 318–329.
Schouten, J. F., and Ornstein, L. S. 1939. Measurements on direct and indirect adaptation by means of a binocular method. *J. Opt. Soc. Amer.*, 19: 168–182.
Semmelroth, C. C. 1970. Predictions of lightness and brightness on different backgrounds. *J. Opt. Soc. Amer.*, 60: 1685–1689.
Sheehan, M.R. 1938. A study of individual consistency in phenomenal constancy. *Arch. Psychol.*, no. 222.
Stevens, S. S. 1934. The attributes of tones. *Proc. Nat. Acad. Sci.*, 20: 457–459.
——. 1961. To honor Fechner and repeal his law. *Science*, 133: 80–86.

——, and Stevens, J. C. 1960. The dynamics of visual brightness. *Psychophysical Project Report Harvard University, no. PPR-246.*

Stewart, E. C. 1959. The Gelb effect. *J. Exp. Psychol.,* 57: 235–242.

Takasaki, H. 1966. Lightness changes of grays induced by change in reflectance of gray background. *J. Opt. Soc. Amer.,* 56: 504–509.

Taubman, R. 1945. Apparent whiteness in relation to albedo and illumination. *J. Exp. Psychol.,* 35: 235–241.

Thouless, R. H. 1931. Phenomenal regression to the "real" object. *Brit. J. Psychol.,* 21: 339–359; 22: 1–30.

——. 1932. Individual differences in phenomenal regression. *Brit. J. Psychol.,* 22: 216–241.

Titchener, E. B. 1910. *A text-book of psychology.* New York: Macmillan.

Torii, S., and Uemura, Y. 1965. Effects of inducing luminance and area upon the apparent brightness of a test field. *Jap. Psychol. Res.,* 2: 86–100.

Treisman, M. 1970. Brightness contrast and the perceptual scale. *Brit. J. Math. and Stat. Psychol.,* 23: 206–224.

Troland, L. T. 1929 (Vol. 1). 1930 (Vol. 2). *The principles of psychophysiology.* New York: Van Nostrand.

Wallach, H. 1948. Brightness constancy and the nature of achromatic colors. *J. Exp. Psychol.,* 38: 310–324.

——. 1963. The perception of neutral colors. *Scientific American,* 208(1): 107–116.

——, and Galloway, A. 1946. The constancy of colored objects in colored illumination. *J. Exp. Psychol.,* 36: 119–126.

——, and O'Connell, D. N. 1950. Perception of tri-dimensional form. *Amer. Psychologist,* 5: 487 (abstract).

Walls, G. L. 1960. Land! Land! *Psychol. Bull.,* 57: 29–48.

Weintraub, D. J. 1964. Successive contrast involving luminance and purity alterations of the Ganzfeld. *J. Exp. Psychol.,* 68: 555–562.

Wheeler, L. 1963. Color matching responses to red light of varying

luminance and purity in complex and simple images. *J. Opt. Soc. Amer.*, 53: 978–993.
Woodworth, R. S. 1938. *Experimental psychology*. New York: Holt.
——, and Schlosberg, H. 1954. *Experimental psychology*. New York: Holt.
Wright, W. D. 1944. *The measurement of colour*. London: Adam Hilger.
——. 1967. *The rays are not coloured*. London: Adam Hilger.
Zweig, H. J. 1956. Autocorrelation and granularity. Part I. Theory. *J. Opt. Soc. Amer.*, 46: 805–811.

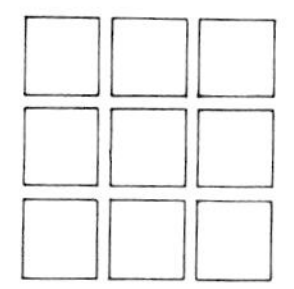

Index

SURFACE COLOR PERCEPTION

Designed by R. E. Rosenbaum.
Composed by Vail-Ballou Press, Inc.,
in 11 point linofilm Baskerville, 2 points leaded,
with display lines in Helvetica.
Printed offset by Vail-Ballou Press
on Warren's Patina II, 60 pound basis.
Bound by Vail-Ballou Press
in Interlaken book cloth
and stamped in All Purpose foil.

Library of Congress Cataloging in Publication Data

(For library cataloging purposes only)
Beck, Jacob.
Surface color perception.

Bibliography: p.
1. Color sense. I Title.
QP481.B34 612·.84 76-38118
ISBN 0-8014-070–4